PHARMACEUTICAL MICROBIOLOGY

PHARMACEUTICAL MICROBIOLOGY

Oleg Paliy
Katherine Bauer

Kruger Brentt
Publishers
2025

Kruger Brentt Publishers UK. LTD.
Company Number 9728962

Regd. Office: 68 St Margarets Road, Edgware, Middlesex HA8 9UU

ISBN: 978-1-78715-446-9

Disclaimer:

Every possible effort has been made to ensure that the information contained in this book is accurate at the time of going to press, and the publisher and author cannot accept responsibility for any errors or omissions, however caused. No responsibility for loss or damage occasioned to any person acting, or refraining from action, as a result of the material in this publication can be accepted by the editor, the publisher or the author. The Publisher is not associated with any product or vendor mentioned in the book. The contents of this work are intended to further general scientific research, understanding and discussion only. Readers should consult with a specialist where appropriate.

Every effort has been made to trace the owners of copyright material used in this book, if any. The author and the publisher will be grateful for any omission brought to their notice for acknowledgement in the future editions of the book.

For information on all our publications visit our website at http://krugerbrentt.com/

PREFACE

Pharmaceutical microbiology is a specialized branch of microbiology that focuses on the study of microorganisms in the context of pharmaceutical products, drug development, and production.

This field plays a critical role in ensuring the safety, efficacy, and quality of pharmaceutical products by preventing contamination, detecting microbial infections, and monitoring microbial resistance. It encompasses various aspects of microbiology, including the identification, isolation, and characterization of microorganisms, as well as the study of their interactions with pharmaceutical formulations.

A key area within pharmaceutical microbiology is the control of microbial contamination in drug manufacturing. Contamination can occur at any stage of the production process, from raw material sourcing to packaging, and can result in compromised product quality, efficacy, or safety. This necessitates stringent quality control measures, such as microbial testing, environmental monitoring, and adherence to aseptic techniques.

Pharmaceutical microbiologists are involved in implementing and validating sterilization and disinfection procedures to eliminate harmful microorganisms, such as bacteria, fungi, and viruses, from the production environment and pharmaceutical products.

Another essential component of pharmaceutical microbiology is the development of antimicrobial agents, including antibiotics and antifungals. Understanding the mechanisms by which microorganisms resist these drugs is fundamental to the development of new therapeutic options.

The study of microbial resistance is a growing concern in pharmaceutical microbiology, as pathogens increasingly develop resistance to existing drugs, leading to the need for innovative treatments.

Pharmaceutical microbiology also plays a significant role in ensuring that vaccines, biologics, and other biotechnology-based products are free from microbial contamination. The development and manufacturing of these products require the use of cell cultures and other biologically derived materials, which can introduce potential microbiological risks.

As such, microbiologists are responsible for ensuring the sterility and safety of these products through rigorous testing and monitoring.

In addition, pharmaceutical microbiologists collaborate with regulatory bodies to ensure compliance with global standards and regulations, such as Good Manufacturing Practices (GMP) and guidelines from the U.S. Food and Drug Administration (FDA) and the World Health Organization (WHO). These regulations are essential to maintaining the integrity of the pharmaceutical industry and protecting public health by ensuring that all products meet the highest standards of microbial quality.

Oleg Paliy

Katherine Bauer

CONTENTS

PART 1:

INTRODUCING PHARMACEUTICAL MICROBIOLOGY

CHAPTER 1

INTRODUCTION TO PHARMACEUTICAL MICROBIOLOGY

Pharmaceutical microbiology is a specialized branch of microbiology that focuses on the study of microorganisms associated with the production of pharmaceuticals. This field plays a critical role in ensuring the safety, quality, and efficacy of medications, from their manufacturing processes to their final application in patient care. Microbial contamination and its control are key concerns for the pharmaceutical industry, as contamination can lead to compromised product quality, potential health risks for patients, and severe financial and reputational losses for companies.

1.1 THE ROLE OF MICROBIOLOGY IN PHARMACEUTICALS

Pharmaceutical microbiology is essential in safeguarding products from microbial contamination and deterioration. Contaminants in drugs, especially injectables and other sterile products, can lead to severe infections or even fatalities in patients. Microbial contamination can arise from multiple sources, including raw materials, manufacturing equipment, or personnel handling the products. Consequently, stringent microbial quality control measures are put in place at every stage of drug development and manufacturing to prevent contamination, maintain sterility, and ensure product safety.

The discipline also encompasses understanding the effects of antimicrobial agents, preservatives, and sterilization methods on microorganisms. Microbiologists evaluate the effectiveness of different sterilization techniques, such as heat, radiation, and filtration, to ensure they meet industry standards for eliminating potential contaminants. Moreover, by studying the behavior and resistance patterns of microbes in various conditions, scientists in pharmaceutical microbiology are better able to predict potential risks and design appropriate strategies to counteract microbial threats.

1.2 AREAS OF FOCUS IN PHARMACEUTICAL MICROBIOLOGY

Pharmaceutical microbiology covers several key areas that support drug safety and efficacy, including microbial testing, sterilization methods, antimicrobial preservation, and contamination control.

1.2.1 Microbial Testing

Microbial testing is a fundamental aspect of pharmaceutical microbiology. This process involves identifying and quantifying microorganisms in products to ensure they fall within safe limits. Key testing methods include the assessment of bioburden, which measures the number of viable microorganisms present in a sample, and the sterility test, which checks for the presence of any living microorganisms in a product. Endotoxin testing, another critical test, screens for bacterial toxins that may be present in drug products and which can cause harmful reactions in patients.

Advanced techniques, such as polymerase chain reaction (PCR) and MALDI-TOF mass spectrometry, are now used in microbial testing to enhance accuracy, sensitivity, and speed in detecting contaminants. These methods allow for the rapid identification of microbial strains, enabling manufacturers to take swift corrective actions when needed.

1.2.2 Sterilization Methods

Sterilization is the process of eliminating all forms of microbial life, including bacteria, fungi, and spores, from pharmaceutical products and equipment. Several sterilization methods are employed in pharmaceutical settings, including autoclaving, dry heat, filtration, gas sterilization (ethylene oxide), and irradiation. Each method has its specific applications and limitations. For instance, heat sterilization is effective but may not be suitable for heat-sensitive products, while filtration can only be used for liquids and cannot remove all viruses.

Pharmaceutical microbiologists play a vital role in determining the most appropriate sterilization technique for each product. The choice depends on factors such as the product's physical and chemical properties, the type of microorganism to be eliminated, and the intended use of the product. Additionally, these professionals evaluate the effectiveness of sterilization methods by conducting sterility assurance tests to confirm the absence of viable microorganisms.

1.2.3 Antimicrobial Preservation

Antimicrobial preservatives are substances added to pharmaceutical products to prevent microbial growth during storage and use. These are especially important in multi-dose products, which can be exposed to microorganisms each time a dose is dispensed. The effectiveness of preservatives is evaluated through a preservative efficacy test (PET), which assesses the ability of a preservative to inhibit microbial growth over time.

The choice of preservatives is critical, as it must not interfere with the drug's efficacy or pose a safety risk to the patient. Pharmaceutical microbiologists determine the optimal type and concentration of preservatives needed to ensure product stability and safety.

1.2.4 Contamination Control

Contamination control is one of the most critical aspects of pharmaceutical microbiology. Contamination in a manufacturing environment can compromise product quality and lead to costly recalls or regulatory actions. To prevent contamination, strict aseptic techniques are applied, especially in facilities that produce sterile products. These techniques include maintaining cleanroom environments, following good manufacturing practices (GMP), and employing proper handling and packaging procedures.

Personnel training, air quality monitoring, and surface testing are essential elements of contamination control. Pharmaceutical microbiologists work closely with quality assurance teams to ensure that microbial contamination risks are minimized, and they are responsible for investigating any contamination events that may occur during production.

1.3 IMPORTANCE OF QUALITY CONTROL AND REGULATORY COMPLIANCE

Pharmaceutical microbiology is closely tied to quality control and regulatory compliance. Regulatory agencies such as the Food and Drug Administration (FDA) in the United States and the European Medicines Agency (EMA) in Europe establish stringent guidelines and standards for microbial testing, sterilization, and contamination control in pharmaceutical manufacturing. Adherence to these regulations is crucial for ensuring the safety and efficacy of drugs on the market.

Pharmaceutical companies are required to validate their manufacturing processes, demonstrate sterility in product testing, and maintain comprehensive documentation to meet regulatory requirements. Pharmaceutical microbiologists ensure that these standards are met by conducting regular tests and monitoring all aspects of the manufacturing environment. Any deviations from regulatory standards can result in penalties or suspension of manufacturing activities.

1.4 CHALLENGES IN PHARMACEUTICAL MICROBIOLOGY

Despite advancements in microbial detection and control, pharmaceutical microbiology faces ongoing challenges. Microorganisms continuously evolve, developing resistance to antimicrobial agents and sterilization methods. Additionally, as pharmaceutical products become more complex, such as biologics and cell-based therapies, the need for specialized microbial testing methods has increased. Ensuring the safety of these advanced therapies poses unique challenges, as traditional microbial testing and sterilization methods may not be suitable.

Therefore, pharmaceutical microbiologists must develop new testing techniques and adapt sterilization methods to address these emerging needs. Moreover, the increase in global manufacturing and outsourcing has added complexity to contamination control, requiring strict oversight to ensure compliance with quality standards across different locations.

1.5 FUTURE OF PHARMACEUTICAL MICROBIOLOGY

The future of pharmaceutical microbiology lies in the adoption of rapid microbial methods, the use of predictive microbiology, and the implementation of advanced technology. Rapid methods, such as real-time PCR and automation in microbial testing, can significantly reduce testing times and improve the detection of contaminants. Predictive microbiology, which involves modeling the growth and behavior of microorganisms, may help in anticipating and preventing contamination issues.

Advances in artificial intelligence and machine learning hold promise for pharmaceutical microbiology, as these technologies can help in analyzing large datasets, identifying trends, and making more accurate predictions about contamination risks. Additionally, as personalized medicine and biotechnology grow, pharmaceutical microbiology will need to develop more precise and targeted microbial testing methods tailored to individual therapies.

1.6 CONCLUSION

Pharmaceutical microbiology is a vital field that ensures the safety, efficacy, and quality of pharmaceutical products. By focusing on microbial testing, sterilization methods, antimicrobial preservation, and contamination control, this discipline supports the pharmaceutical industry in its mission to produce safe and effective drugs for patients worldwide. However, the field is not without challenges, as microbial resistance, complex drug formulations, and global manufacturing practices continue to pose risks to product safety. Looking forward, innovations in rapid microbial detection, predictive microbiology, and artificial intelligence hold the potential to transform pharmaceutical microbiology, enabling it to keep pace with the evolving needs of modern medicine. As the industry progresses, pharmaceutical microbiologists will continue to play a crucial role in ensuring that products meet the highest standards of quality and safety, safeguarding public health in the process.

PART 2:

BIOLOGY OF MICROORGANISMS

CHAPTER 2

FUNDAMENTAL FEATURES OF MICROBIOLOGY

Microbiology is the branch of biology that studies microorganisms, which are microscopic, single-celled organisms such as bacteria, viruses, fungi, protozoa, and algae.

Although invisible to the naked eye, microorganisms are fundamental to life on Earth, influencing ecosystems, human health, and various industries. The study of microbiology encompasses a wide range of topics, from understanding basic microbial structures and functions to exploring their roles in nature and human applications.

2.1 MICROBIAL DIVERSITY: EXPLORING THE MICROBIAL WORLD

Microbial diversity encompasses the vast array of microorganisms that inhabit nearly every environment on Earth, from soil and water to the human body and extreme environments like hot springs and deep-sea vents.

These organisms include bacteria, archaea, fungi, algae, viruses, and protozoa, each of which plays unique roles in their ecosystems. Understanding microbial diversity is essential for fields such as medicine, agriculture, environmental science, and biotechnology, as these tiny organisms impact life on a macroscopic scale.

Microbial diversity is a critical component of life on Earth, impacting everything from ecosystem functioning to human health and industrial applications. The variety of microorganisms and their adaptive capabilities enable them to survive in almost any environment, highlighting their resilience and ecological importance. By studying and conserving microbial diversity, we gain valuable insights into ecosystem processes, enhance agricultural productivity, and develop biotechnological solutions to global challenges.

2.1.1 The Types of Microorganisms: Key Players in the Microbial World

Microbial diversity can be broadly categorized into groups based on cellular structure, function, and ecological roles. The major types of microorganisms include bacteria, archaea, fungi, algae, protozoa, and viruses.

- **Bacteria:** Bacteria are single-celled prokaryotes with diverse shapes, including spherical, rod-shaped, and spiral. They inhabit a variety of environments and exhibit a wide range of metabolic capabilities. Some bacteria are pathogens, causing diseases, while others play beneficial roles in processes like nitrogen fixation, digestion, and decomposition. They are among the oldest life forms, having adapted to survive in almost every environment on Earth.
- **Archaea:** Similar to bacteria in shape and size, archaea are also prokaryotes but are genetically distinct from bacteria. They often inhabit extreme environments, such as acidic hot springs, high-salinity waters, and anaerobic environments. Despite these extreme habitats, archaea play important roles in ecosystems, particularly in methane production and sulfur cycling.
- **Fungi:** Fungi include yeasts, molds, and mushrooms, all of which are eukaryotic organisms with complex cell structures. Fungi are key decomposers in the environment, breaking down organic matter into simpler compounds, which recycle nutrients back into the ecosystem. Some fungi form symbiotic relationships with plants (mycorrhizae), aiding in nutrient absorption, while others are pathogens that can infect plants, animals, and humans.
- **Algae:** Algae are photosynthetic eukaryotes found in freshwater and marine environments. They produce oxygen and form the base of aquatic food chains, supporting diverse ecosystems. Algae come in various forms, including single-celled diatoms, multicellular seaweeds, and complex forms like kelp. Through photosynthesis, algae play a significant role in carbon dioxide sequestration, contributing to the regulation of atmospheric gases.
- **Protozoa:** Protozoa are single-celled eukaryotes found in water and soil. They are generally motile and feed on bacteria, organic matter, and other small organisms. Protozoa contribute to the regulation of bacterial populations in ecosystems and play roles in nutrient cycling. However, some protozoa are parasites that cause diseases in humans, animals, and plants.
- **Viruses:** Viruses are unique microorganisms that lack cellular structures and require a host cell to replicate. They infect all forms of life, from humans and animals to plants and bacteria. While some viruses cause diseases, others have beneficial effects, such as influencing microbial populations and genetic diversity. Phages, or viruses that infect bacteria, are also used in research to target harmful bacterial pathogens.

2.1.2 Functions and Roles of Microorganisms in Ecosystems

Microorganisms play fundamental roles in ecosystems, contributing to nutrient cycling, primary production, and the breakdown of organic materials. Their roles are diverse and vary according to their metabolic capabilities and environmental requirements.

- **Nutrient Cycling:** Microorganisms are crucial to nutrient cycles, including the carbon, nitrogen, and sulfur cycles. For example, nitrogen-fixing bacteria convert atmospheric nitrogen into forms that plants can use, making it available to other organisms in the ecosystem. Similarly, sulfur-reducing bacteria help in breaking down sulfur compounds, making sulfur accessible to plants and animals.
- **Primary Production:** Photosynthetic microorganisms, such as algae and cyanobacteria, contribute to primary production in aquatic ecosystems by converting solar energy into chemical energy, which is then consumed by other organisms in the food web. This process not only provides energy but also releases oxygen as a byproduct, which supports aerobic life forms.
- **Decomposition:** Decomposer microorganisms, such as fungi and certain bacteria, break down dead organic matter, returning nutrients to the soil and helping maintain ecosystem health. Without decomposers, organic material would accumulate, leading to nutrient imbalances.
- **Symbiotic Relationships:** Microorganisms form symbiotic relationships with plants, animals, and even humans. For example, mycorrhizal fungi associate with plant roots, helping them absorb nutrients in exchange for carbohydrates. In humans, the gut microbiome supports digestion, immunity, and even mental health. These relationships are crucial for the survival and health of various species, highlighting the importance of microbial diversity.

2.1.3 The Importance of Microbial Diversity

Microbial diversity is essential for ecosystem stability, resilience, and productivity. Diverse microbial communities contribute to ecosystem functions, such as nutrient cycling, and provide ecosystem services, including water purification and soil fertility.

- **Ecosystem Stability and Resilience:** Diverse microbial communities enhance ecosystem stability by increasing redundancy. If one microbial species declines, others with similar functions can compensate, ensuring that essential ecosystem processes continue. This resilience is particularly important in the face of environmental changes, such as pollution or climate change.

- **Human Health and Agriculture:** In agriculture, microbial diversity is essential for soil health and crop productivity. A diverse soil microbiome supports plant growth by enhancing nutrient availability and suppressing pathogens. In human health, the diversity of the gut microbiome has been linked to overall well-being, with imbalances potentially leading to various health issues, including obesity, allergies, and autoimmune diseases.
- **Biotechnology and Industry:** Microorganisms are valuable in biotechnology, where they are used to produce antibiotics, biofuels, and enzymes. Diverse microbial communities provide a rich source of genetic material that can be harnessed for various applications, from medicine to environmental cleanup.

2.1.4 Factors Influencing Microbial Diversity

Several factors influence microbial diversity, including environmental conditions, nutrient availability, and interactions with other organisms. These factors shape microbial communities in specific ecosystems.

- **Temperature and pH:** Microorganisms have specific temperature and pH ranges in which they thrive. For instance, thermophilic bacteria prefer hot environments, while acidophiles can survive in acidic conditions. Temperature and pH influence not only survival but also metabolic rates, affecting how microorganisms interact within their ecosystems.
- **Nutrient Availability:** Microbial diversity is influenced by the availability of nutrients, as microorganisms require specific nutrients for growth. In nutrient-rich environments, microbial populations grow rapidly, while in nutrient-poor settings, only specialized organisms with unique metabolic capabilities can survive. This variation contributes to the diversity of microbial communities in different environments.
- **Interactions with Other Organisms:** Microorganisms interact with other species, both microbial and macroscopic, in various ways, including competition, predation, and symbiosis. These interactions shape microbial diversity and influence ecosystem dynamics. For example, protozoa consume bacteria, regulating bacterial populations, while mutualistic relationships with plants allow certain microbes to thrive in plant root zones.

2.1.5 Microbial Diversity and Environmental Adaptation

Microorganisms have evolved to adapt to extreme and diverse environments, contributing to the concept of extremophiles. Extremophiles are microorganisms that thrive in conditions previously thought to be uninhabitable, such as high-salt, high-temperature, and high-pressure environments.

- **Adaptation to Extreme Environments:** Extremophiles include halophiles (salt-loving), thermophiles (heat-loving), and psychrophiles (cold-loving). These organisms have adapted unique cellular mechanisms to withstand extreme conditions, such as specialized enzymes and cell membranes. The study of extremophiles has implications for biotechnology and astrobiology, as they provide insights into how life might survive on other planets.
- **Genetic Adaptation and Evolution:** Microbial diversity is driven by genetic diversity, as microorganisms constantly evolve through mutations, gene transfer, and recombination. This genetic adaptability allows microbes to survive in changing environments, which contributes to ecosystem resilience and enables the rapid emergence of traits like antibiotic resistance in pathogenic bacteria.

2.1.6 Conservation of Microbial Diversity

The conservation of microbial diversity is essential for maintaining ecosystem functions and services. Human activities, such as deforestation, pollution, and overuse of antibiotics, threaten microbial diversity. These disturbances can lead to the loss of essential microbial species, impacting ecosystem stability and productivity.

- **Impacts of Human Activity:** Pollution and habitat destruction disrupt microbial communities, leading to reduced diversity and ecosystem degradation. For example, excessive fertilizer use can lead to algal blooms in aquatic systems, disrupting the balance of microbial communities and harming other species.
- **Efforts in Conservation:** Conservation efforts focus on preserving natural environments, reducing pollution, and promoting sustainable agricultural practices to protect microbial diversity. Research in microbial ecology is also expanding, allowing for better understanding and management of microbial ecosystems.

2.2 MICROBIAL STRUCTURE AND FUNCTION: UNDERSTANDING MICROBIAL CELL TYPES

The structural diversity of microorganisms reflects their wide range of functions and adaptability to different environments. In microbiology, microorganisms are classified as either prokaryotes or eukaryotes based on their cellular organization.

2.2.1 Prokaryotes

A prokaryote is a single-cell organism whose cell lacks a nucleus and other membrane-bound organelles. In the two-empire system arising from the work of Édouard Chatton, prokaryotes were classified within the empire Prokaryota. However in the three-domain system, based upon molecular analysis, prokaryotes are divided into two domains: *Bacteria* (formerly Eubacteria) and *Archaea* (formerly Archaebacteria). Organisms with nuclei are placed in a third domain: Eukaryota.

Prokaryotes evolved before eukaryotes, and lack nuclei, mitochondria, and most of the other distinct organelles that characterize the eukaryotic cell. It was once thought that prokaryotic cellular components were unenclosed within the cytoplasm except for an outer cell membrane, but bacterial microcompartments, which are thought to be quasi-organelles enclosed in protein shells (such as the encapsulin protein cages), have been discovered, along with other prokaryotic organelles. While being unicellular, some prokaryotes, such as cyanobacteria, may form colonies held together by biofilms, and large colonies can create multilayered microbial mats. Others, such as myxobacteria, have multicellular stages in their life cycles. Prokaryotes are asexual, reproducing via binary fission without any fusion of gametes, although horizontal gene transfer may take place.

Molecular studies have provided insight into the evolution and interrelationships of the three domains of life. The division between prokaryotes and eukaryotes reflects the existence of two very different levels of cellular organization; only eukaryotic cells have an enveloped nucleus that contains its chromosomal DNA, and other characteristic membrane-bound organelles including mitochondria. Distinctive types of prokaryotes include extremophiles and methanogens; these are common in some extreme environments.

History

The distinction between prokaryotes and eukaryotes was firmly established by the microbiologists Roger Stanier and C. B. van Niel in their 1962 paper *The concept of a bacterium* (though spelled procaryote and eucaryote there). That paper cites Édouard Chatton's 1937 book *Titres et Travaux Scientifiques* for using those terms and recognizing the distinction. One reason for this classification was so that what was then often called blue-green algae (now called cyanobacteria) would not be classified as plants but grouped with bacteria.

Structure

Prokaryotes have a prokaryotic cytoskeleton that is more primitive than that of the eukaryotes. Besides homologues of actin and tubulin (MreB and FtsZ), the helically arranged building-block of the flagellum, flagellin, is one of the most significant cytoskeletal proteins of bacteria, as it provides structural backgrounds of chemotaxis, the basic cell physiological response of bacteria. At least some prokaryotes also contain intracellular structures that can be seen as primitive organelles. Membranous organelles (or intracellular membranes) are known in some groups of prokaryotes, such as vacuoles or membrane systems devoted to special metabolic properties, such as photosynthesis or chemolithotrophy. In addition, some species also contain carbohydrate-enclosed microcompartments, which have distinct physiological roles (e.g. carboxysomes or gas vacuoles).

Most prokaryotes are between 1 μm and 10 μm, but they can vary in size from 0.2 μm (*Mycoplasma genitalium*) to 750 μm (*Thiomargarita namibiensis*).

Prokaryotic Cell Structure	Description
Flagellum (not always present)	Long, whip-like protrusion that aids cellular locomotion used by both gram-positive and gram-negative bacteria.
Cell membrane	Surrounds the cell's cytoplasm and regulates the flow of substances in and out of the cell.
Cell wall (except genera *Mycoplasma* and *Thermoplasma*)	Outer covering of most cells that protects the bacterial cell and gives it shape.
Cytoplasm	A gel-like substance composed mainly of water that also contains enzymes, salts, cell components, and various organic molecules.
Ribosome	Cell structures responsible for protein production.
Nucleoid	Area of the cytoplasm that contains the prokaryote's single DNA molecule.
Glycocalyx (only in some types of prokaryotes)	A glycoprotein-polysaccharide covering that surrounds the cell membranes.
Cytoplasmic inclusions	The inclusions such as ribosomes and larger masses scattered in the cytoplasmic matrix.

Morphology

Prokaryotic cells have various shapes; the four basic shapes of bacteria are:

- **Cocci:** A bacterium that is spherical or ovoid is called a coccus (Plural, cocci)., e.g., *Streptococcus, Staphylococcus.*
- **Bacilli:** A bacterium with cylindrical shape called rod or a bacillus (Plural, bacilli).
- **Spiral Bacteria:** Some rods twist into spiral shapes and are called spirilla (singular, spirillum).
- **Vibrio:** Comma-shaped.

The archaeon Haloquadratum has flat square-shaped cells.

Reproduction

Bacteria and archaea reproduce through asexual reproduction, usually by binary fission. Genetic exchange and recombination still occur, but this is a form of horizontal gene transfer and is not a replicative process, simply involving the transference of DNA between two cells, as in bacterial conjugation.

DNA Transfer

DNA transfer between prokaryotic cells occurs in bacteria and archaea, although it has been mainly studied in bacteria. In bacteria, gene transfer occurs by three processes. These are:

- Bacterial virus (bacteriophage)-mediated transduction.
- Plasmid-mediated conjugation.
- Natural transformation.

Transduction of bacterial genes by bacteriophage appears to reflect an occasional error during intracellular assembly of virus particles, rather than an adaptation of the host bacteria. The transfer of bacterial DNA is under the control of the bacteriophage's genes rather than bacterial genes. Conjugation in the well-studied *E. coli* system is controlled by plasmid genes, and is an adaptation for distributing copies of a plasmid from one bacterial host to another. Infrequently during this process, a plasmid may integrate into the host bacterial chromosome, and subsequently transfer part of the host bacterial DNA to another bacterium. Plasmid mediated transfer of host bacterial DNA (conjugation) also appears to be an accidental process rather than a bacterial adaptation.

Natural bacterial transformation involves the transfer of DNA from one bacterium to another through the intervening medium. Unlike transduction and conjugation, transformation is clearly a bacterial adaptation for DNA transfer, because it depends on numerous bacterial gene products that specifically interact to perform this complex process. For a bacterium to bind, take up and recombine donor DNA into its own chromosome, it must first enter a special physiological state called competence. About 40 genes are required in *Bacillus subtilis* for the development of competence. The length of DNA transferred during *B. subtilis* transformation can be as much as a third to the whole chromosome. Transformation is a common mode of DNA transfer, and 67 prokaryotic species are thus far known to be naturally competent for transformation.

Among archaea, *Halobacterium volcanii* forms cytoplasmic bridges between cells that appear to be used for transfer of DNA from one cell to another. Another archaeon, *Sulfolobus solfataricus*, transfers DNA between cells by direct contact. Frols et al. (2008) found that exposure of *S. solfataricus* to DNA damaging agents induces cellular aggregation, and suggested that cellular aggregation may enhance DNA transfer among cells to provide increased repair of damaged DNA via homologous recombination.

Sociality

While prokaryotes are considered strictly unicellular, most can form stable aggregate communities. When such communities are encased in a stabilizing polymer matrix ("slime"), they may be called "biofilms". Cells in biofilms often show distinct patterns of gene expression (phenotypic differentiation) in time and space. Also, as with multicellular eukaryotes, these changes in expression often appear to result from cell-to-cell signaling, a phenomenon known as quorum sensing.

Biofilms may be highly heterogeneous and structurally complex and may attach to solid surfaces, or exist at liquid-air interfaces, or potentially even liquid-liquid interfaces. Bacterial biofilms are often made up of microcolonies (approximately dome-shaped masses of bacteria and matrix) separated by "voids" through which the medium (e.g., water) may flow easily.

The microcolonies may join together above the substratum to form a continuous layer, closing the network of channels separating microcolonies. This structural complexity—combined with observations that oxygen limitation (a ubiquitous challenge for anything growing in size beyond the scale of diffusion) is at least partially eased by movement of medium throughout the biofilm—has led some to speculate that this may constitute a circulatory system and many researchers have started calling prokaryotic communities multicellular (for example). Differential cell expression, collective behavior, signaling, programmed cell death, and (in some cases) discrete biological dispersal events all seem to point in this direction.

However, these colonies are seldom if ever founded by a single founder (in the way that animals and plants are founded by single cells), which presents a number of theoretical issues. Most explanations of co-operation and the evolution of multicellularity have focused on high relatedness between members of a group (or colony, or whole organism).

If a copy of a gene is present in all members of a group, behaviors that promote cooperation between members may permit those members to have (on average) greater fitness than a similar group of selfish individuals.

Should these instances of prokaryotic sociality prove to be the rule rather than the exception, it would have serious implications for the way we view prokaryotes in general, and the way we deal with them in medicine. Bacterial biofilms may be 100 times more resistant to antibiotics than free-living unicells and may be nearly impossible to remove from surfaces once they have colonized them. Other aspects of bacterial cooperation—such as bacterial conjugation and quorum-sensing-mediated pathogenicity, present additional challenges to researchers and medical professionals seeking to treat the associated diseases.

Environment

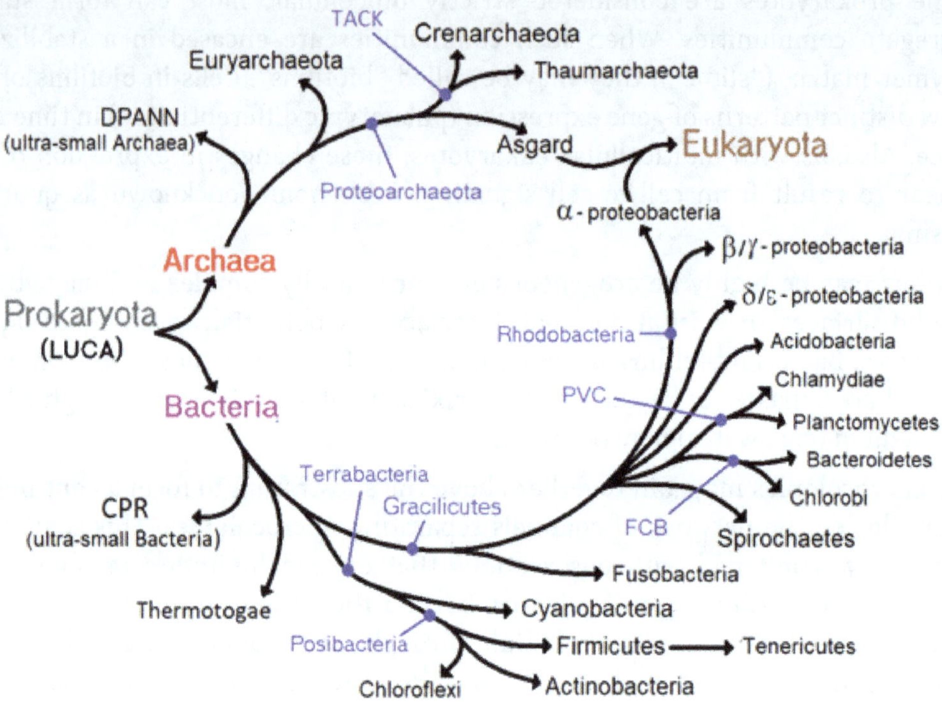

Fig 1: Phylogenetic Ring Showing the Diversity of Prokaryotes, and Symbiogenetic Origins of Eukaryotes

Prokaryotes have diversified greatly throughout their long existence. The metabolism of prokaryotes is far more varied than that of eukaryotes, leading to many highly distinct prokaryotic types. For example, in addition to using photosynthesis or organic compounds for energy, as eukaryotes do, prokaryotes may obtain energy from inorganic compounds such as hydrogen sulfide. This enables prokaryotes to thrive in harsh environments as cold as the snow surface of Antarctica, studied in cryobiology, or as hot as undersea hydrothermal vents and land-based hot springs.

Prokaryotes live in nearly all environments on Earth. Some archaea and bacteria are extremophiles, thriving in harsh conditions, such as high temperatures (thermophiles) or high salinity (halophiles). Many archaea grow as plankton in the oceans. Symbiotic prokaryotes live in or on the bodies of other organisms, including humans. Prokaryote have high populations in the soil - including the rhizosphere and rhizosheath. Soil prokaryotes are still heavily undercharacterized despite their easy proximity to humans and their tremendous economic importance to agriculture.

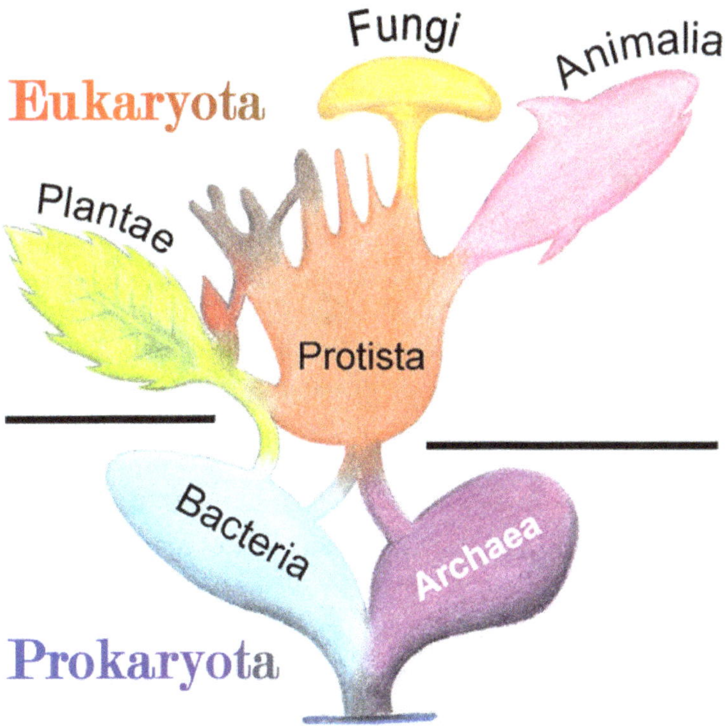

Fig 2: Phylogenetic and Symbiogenetic Tree of Living Organisms, Showing the Origins of Eukaryotes and Prokaryotes

Classification

In 1977, Carl Woese proposed dividing prokaryotes into the Bacteria and Archaea (originally Eubacteria and Archaebacteria) because of the major differences in the structure and genetics between the two groups of organisms. Archaea were originally thought to be extremophiles, living only in inhospitable conditions such as extremes of temperature, pH, and radiation but have since been found in all types of habitats. The resulting arrangement of Eukaryota (also called "Eucarya"), Bacteria, and Archaea is called the three-domain system, replacing the traditional two-empire system.

Evolution

A widespread current model of the evolution of the first living organisms is that these were some form of prokaryotes, which may have evolved out of protocells, while the eukaryotes evolved later in the history of life. Some authors have questioned this conclusion, arguing that the current set of prokaryotic species may have evolved from more complex eukaryotic ancestors through a process of simplification.

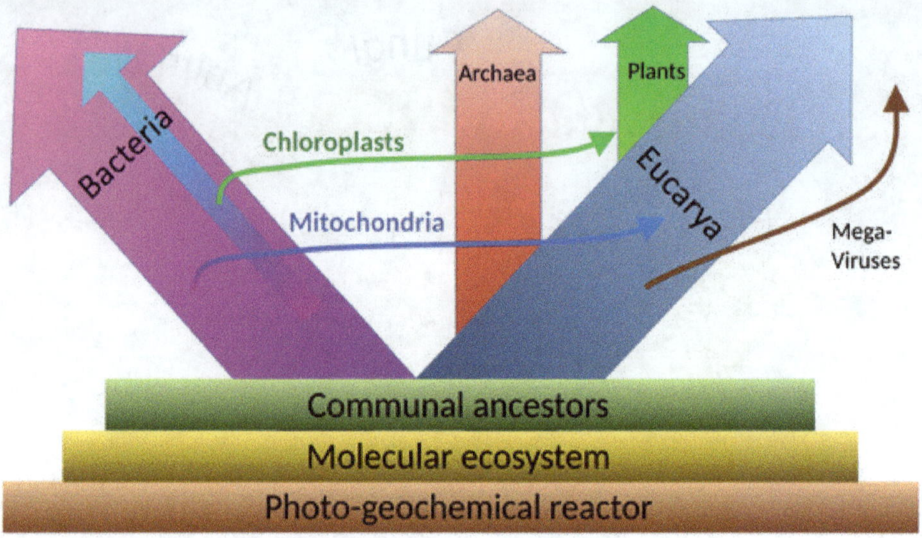

Fig 3: Diagram of the Origin of Life with the Eukaryotes Appearing Early, not Derived from Prokaryotes

Others have argued that the three domains of life arose simultaneously, from a set of varied cells that formed a single gene pool. This controversy was summarized in 2005:

- There is no consensus among biologists concerning the position of the eukaryotes in the overall scheme of cell evolution. Current opinions on the origin and position of eukaryotes span a broad spectrum including the views that eukaryotes arose first in evolution and that prokaryotes descend from them, that eukaryotes arose contemporaneously with eubacteria and archaebacteria and hence represent a primary line of descent of equal age and rank as the prokaryotes, that eukaryotes arose through a symbiotic event entailing an endosymbiotic origin of the nucleus, that eukaryotes arose without endosymbiosis, and that eukaryotes arose through a symbiotic event entailing a simultaneous endosymbiotic origin of the flagellum and the nucleus, in addition to many other models, which have been reviewed and summarized elsewhere.

The oldest known fossilized prokaryotes were laid down approximately 3.5 billion years ago, only about 1 billion years after the formation of the Earth's crust. Eukaryotes only appear in the fossil record later, and may have formed from endosymbiosis of multiple prokaryote ancestors. The oldest known fossil eukaryotes are about 1.7 billion years old. However, some genetic evidence suggests eukaryotes appeared as early as 3 billion years ago. While Earth is the only place in the universe where life is known to exist, some have suggested that there is evidence on Mars of fossil or living prokaryotes. However, this possibility remains the subject of considerable debate and skepticism.

Relationship to Eukaryotes

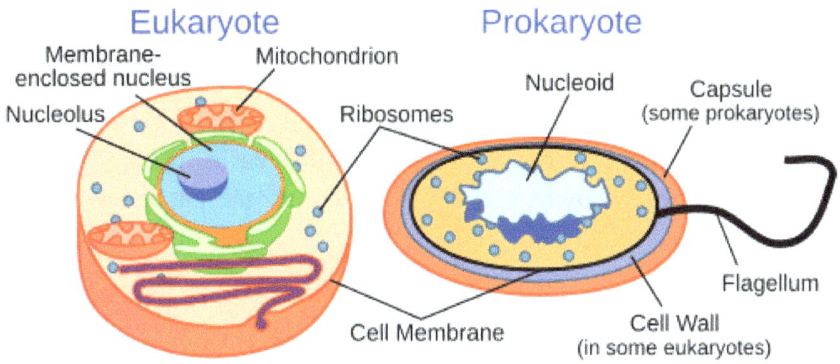

Fig 4: Comparison of Eukaryotes vs. Prokaryotes

The division between prokaryotes and eukaryotes is usually considered the most important distinction or difference among organisms. The distinction is that eukaryotic cells have a "true" nucleus containing their DNA, whereas prokaryotic cells do not have a nucleus.

Both eukaryotes and prokaryotes contain large RNA/protein structures called ribosomes, which produce protein, but the ribosomes of prokaryotes are smaller than those of eukaryotes. Mitochondria and chloroplasts, two organelles found in many eukaryotic cells, contain ribosomes similar in size and makeup to those found in prokaryotes. This is one of many pieces of evidence that mitochondria and chloroplasts are descended from free-living bacteria. The endosymbiotic theory holds that early eukaryotic cells took in primitive prokaryotic cells by phagocytosis and adapted themselves to incorporate their structures, leading to the mitochondria and chloroplasts.

The genome in a prokaryote is held within a DNA/protein complex in the cytosol called the nucleoid, which lacks a nuclear envelope. The complex contains a single, cyclic, double-stranded molecule of stable chromosomal DNA, in contrast to the multiple linear, compact, highly organized chromosomes found in eukaryotic cells. In addition, many important genes of prokaryotes are stored in separate circular DNA structures called plasmids.

Like Eukaryotes, prokaryotes may partially duplicate genetic material, and can have a haploid chromosomal composition that is partially replicated, a condition known as merodiploidy.

Prokaryotes lack mitochondria and chloroplasts. Instead, processes such as oxidative phosphorylation and photosynthesis take place across the prokaryotic cell membrane. However, prokaryotes do possess some internal structures, such as prokaryotic cytoskeletons. It has been suggested that the bacterial phylum Planctomycetota has a membrane around the nucleoid and contains other membrane-bound cellular structures. However, further investigation revealed that Planctomycetota cells are not compartmentalized or nucleated and, like other bacterial membrane systems, are interconnected.

Prokaryotic cells are usually much smaller than eukaryotic cells. Therefore, prokaryotes have a larger surface-area-to-volume ratio, giving them a higher metabolic rate, a higher growth rate, and as a consequence, a shorter generation time than eukaryotes.

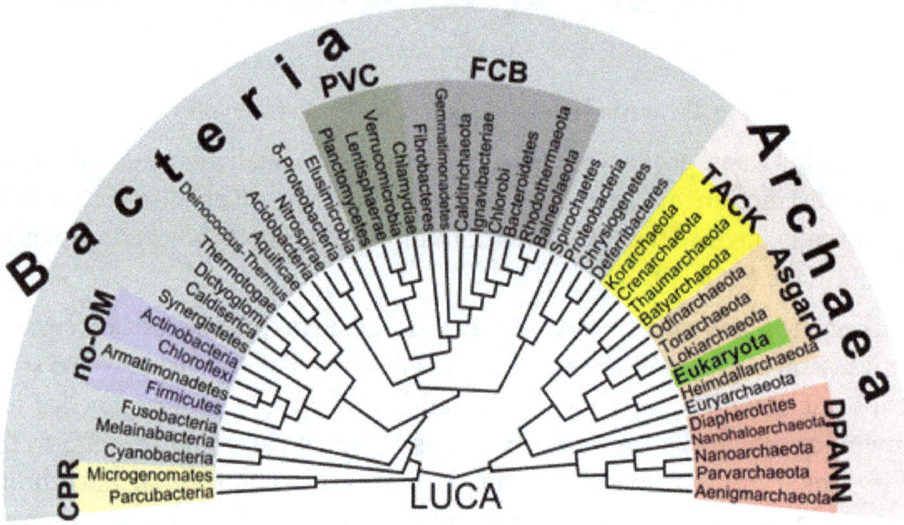

Fig 5: Phylogenetic Tree Showing the Diversity of Prokaryotes

This 2018 proposal shows eukaryotes emerging from the archaean Asgard group which represents a modern version of the eocyte hypothesis. Unlike earlier assumptions, the division between bacteria and the rest is the most important difference between organisms.

There is increasing evidence that the roots of the eukaryotes are to be found in (or at least next to) the archaean asgard group, perhaps Heimdallarchaeota (an idea which is a modern version of the 1984 eocyte hypothesis, *eocytes* being an old synonym for *Thermoproteota*, a taxon to be found nearby the then-unknown Asgard group). For example, histones which usually package DNA in eukaryotic nuclei, have also been found in several archaean groups, giving evidence for homology. This idea might clarify the mysterious predecessor of eukaryotic cells (eucytes) which engulfed an alphaproteobacterium forming the first eucyte (LECA, last eukaryotic common ancestor) according to endosymbiotic theory. There might have been some additional support by viruses, called viral eukaryogenesis. The non-bacterial group comprising archaea and eukaryota was called Neomura by Thomas Cavalier-Smith in 2002. However, in a cladistic view, eukaryota *are* archaea in the same sense as birds *are* dinosaurs because they evolved from the maniraptora dinosaur group. In contrast, archaea *without* eukaryota appear to be a paraphyletic group, just like dinosaurs without birds.

Prokaryotes may be Split into Two Groups

Unlike the assumption of a fundamental split between prokaryotes and eukaryotes, the most important difference between biota may be the division between bacteria and the rest (archaea and eukaryota). For instance, DNA replication differs fundamentally between bacteria and archaea (including that in eukaryotic nuclei), and it may not be homologous between these two groups. Moreover, ATP synthase, though common (homologous) in all organisms, differs greatly between bacteria (including eukaryotic organelles such as mitochondria and chloroplasts) and the archaea/eukaryote nucleus group. The last common antecessor of all life (called LUCA, last universal common ancestor) should have possessed an early version of this protein complex. As ATP synthase is obligate membrane bound, this supports the assumption that LUCA was a cellular organism. The RNA world hypothesis might clarify this scenario, as LUCA might have been a ribocyte (also called ribocell) lacking DNA, but with an RNA genome built by ribosomes as primordial self-replicating entities.

A Peptide-RNA world (also called RNP world) hypothesis has been proposed based on the idea that oligopeptides may have been built together with primordial nucleic acids at the same time, which also supports the concept of a ribocyte as LUCA. The feature of DNA as the material base of the genome might have then been adopted separately in bacteria and in archaea (and later eukaryote nuclei), presumably by help of some viruses (possibly retroviruses as they could reverse transcribe RNA to DNA). As a result, prokaryota comprising bacteria and archaea may also be polyphyletic.

2.2.2 Eukaryotes

The eukaryotes constitute the domain of Eukaryota or Eukarya, organisms whose cells have a membrane-bound nucleus. All animals, plants, fungi, and many unicellular organisms are eukaryotes. They constitute a major group of life forms alongside the two groups of prokaryotes: the Bacteria and the Archaea. Eukaryotes represent a small minority of the number of organisms, but given their generally much larger size, their collective global biomass is much larger than that of prokaryotes.

The eukaryotes seemingly emerged within the Asgard archaea, and are closely related to the Heimdallarchaeia. This implies that there are only two domains of life, Bacteria and Archaea, with eukaryotes incorporated among the Archaea. Eukaryotes first emerged during the Paleoproterozoic, likely as flagellated cells.

The leading evolutionary theory is they were created by symbiogenesis between an anaerobic Asgard archaean and an aerobic proteobacterium, which formed the mitochondria. A second episode of symbiogenesis with a cyanobacterium created the plants, with chloroplasts.

Eukaryotic cells contain membrane-bound organelles such as the nucleus, the endoplasmic reticulum, and the Golgi apparatus. Eukaryotes may be either unicellular or multicellular. In comparison, prokaryotes are typically unicellular. Unicellular eukaryotes are sometimes called protists. Eukaryotes can reproduce both asexually through mitosis and sexually through meiosis and gamete fusion (fertilization).

History of Classification

In antiquity, the two lineages of animals and plants were recognized by Aristotle and Theophrastus. The lineages were given the taxonomic rank of Kingdom by Linnaeus in the 18th century. Though he included the fungi with plants with some reservations, it was later realized that they are quite distinct and warrant a separate kingdom.

The various single-cell eukaryotes were originally placed with plants or animals when they became known. In 1818, the German biologist Georg A. Goldfuss coined the word *protozoa* to refer to organisms such as ciliates, and this group was expanded until Ernst Haeckel made it a kingdom encompassing all single-celled eukaryotes, the Protista, in 1866. The eukaryotes thus came to be seen as four kingdoms:

- Kingdom Protista.
- Kingdom Plantae.
- Kingdom Fungi.
- Kingdom Animalia.

The protists were at that time thought to be "primitive forms", and thus an evolutionary grade, united by their primitive unicellular nature. Understanding of the oldest branchings in the tree of life only developed substantially with DNA sequencing, leading to a system of domains rather than kingdoms as top level rank being put forward by Carl Woese, Otto Kandler, and Mark Wheelis in 1990, uniting all the eukaryote kingdoms in the domain "Eucarya", stating, however, that "'eukaryotes' will continue to be an acceptable common synonym". In 1996, the evolutionary biologist Lynn Margulis proposed to replace Kingdoms and

Domains with "inclusive" names to create a "symbiosis-based phylogeny", giving the description "Eukarya (symbiosis-derived nucleated organisms)".

Origin of Eukaryotes

The origin of the eukaryotic cell, or *eukaryogenesis*, is a milestone in the evolution of life, since eukaryotes include all complex cells and almost all multicellular organisms. The last eukaryotic common ancestor (LECA) is the hypothetical origin of all living eukaryotes, and was most likely a biological population, not a single individual. The LECA is believed to have been a protist with a nucleus, at least one centriole and flagellum, facultatively aerobic mitochondria, sex (meiosis and syngamy), a dormant cyst with a cell wall of chitin or cellulose, and peroxisomes.

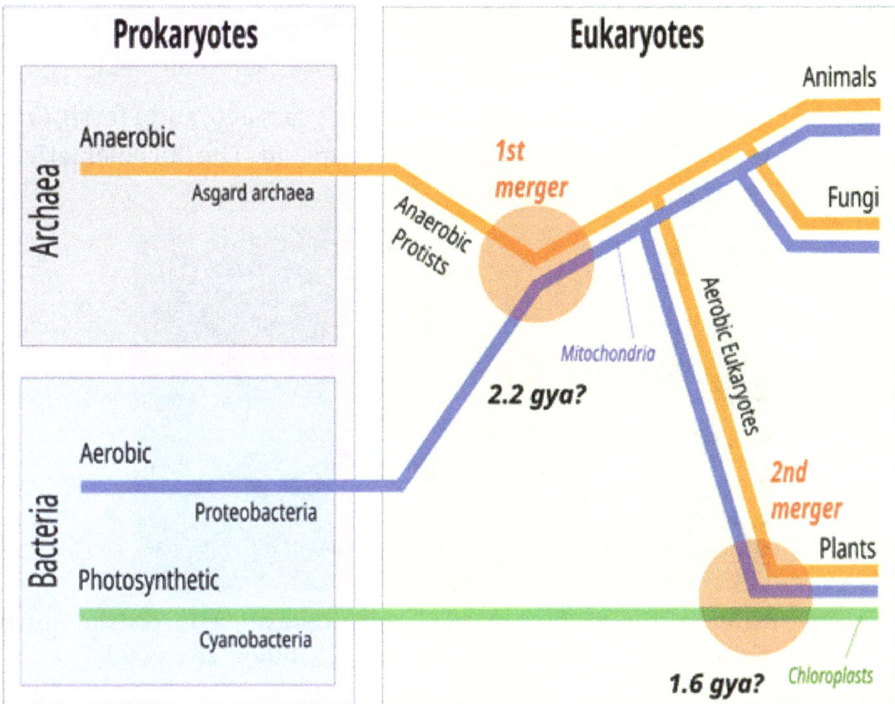

Fig 6: In the Theory of Symbiogenesis, a Merger of an Archaean and an Aerobic Bacterium Created the Eukaryotes, with Aerobic Mitochondria; a Second Merger Added Chloroplasts, Creating the Green Plants

An endosymbiotic union between a motile anaerobic archaean and an aerobic alphaproteobacterium gave rise to the LECA and all eukaryotes, with mitochondria. A second, much later endosymbiosis with a cyanobacterium gave rise to the ancestor of plants, with chloroplasts.

The presence of eukaryotic biomarkers in archaea points towards an archaeal origin. The genomes of Asgard archaea have plenty of Eukaryotic signature protein genes, which play a crucial role in the development of the cytoskeleton and complex cellular structures characteristic of eukaryotes. In 2022, cryo-electron tomography demonstrated that Asgard archaea have a complex actin-based cytoskeleton, providing the first direct visual evidence of the archaeal ancestry of eukaryotes.

Fossils

The timing of the origin of eukaryotes is hard to determine but the discovery of *Qingshania magnificia*, the earliest multicelluar eukaryote from North China which lived during 1.635 billion years ago, suggests that the crown group eukaryotes would have originated from the late Paleoproterozoic (Statherian); the earliest unequivocal unicellular eukaryotes which lived during approximately 1.65 billion years ago are also discovered from North China: *Tappania plana*, *Shuiyousphaeridium macroreticulatum*, *Dictyosphaera macroreticulata*, *Germinosphaera alveolata*, and *Valeria lophostriata*.

Some acritarchs are known from at least 1.65 billion years ago, and a fossil, *Grypania*, which may be an alga, is as much as 2.1 billion years old. The "problematic" fossil *Diskagma* has been found in paleosols 2.2 billion years old.

Fig 7: Reconstruction of the Problematic *Diskagma Buttonii*, **a Terrestrial Fossil Less than 1mm High, from Rocks Around 2.2 Billion Years Old**

Structures proposed to represent "large colonial organisms" have been found in the black shales of the Palaeoproterozoic such as the Francevillian B Formation, in Gabon, dubbed the "Francevillian biota" which is dated at 2.1 billion years old. However, the status of these structures as fossils is contested, with other authors suggesting that they might represent pseudofossils.

The oldest fossils than can unambiguously be assigned to eukaryotes are from the Ruyang Group of China, dating to approximately 1.8-1.6 billion years ago. Fossils that are clearly related to modern groups start appearing an estimated 1.2 billion years ago, in the form of red algae, though recent work suggests the existence of fossilized filamentous algae in the Vindhya basin dating back perhaps to 1.6 to 1.7 billion years ago.

The presence of steranes, eukaryotic-specific biomarkers, in Australian shales previously indicated that eukaryotes were present in these rocks dated at 2.7 billion years old, but these Archaean biomarkers have been rebutted as later contaminants. The oldest valid biomarker records are only around 800 million years old. In contrast, a molecular clock analysis suggests the emergence of sterol biosynthesis as early as 2.3 billion years ago. The nature of steranes as eukaryotic biomarkers is further complicated by the production of sterols by some bacteria.

Whenever their origins, eukaryotes may not have become ecologically dominant until much later; a massive increase in the zinc composition of marine sediments 800 million years ago has been attributed to the rise of substantial populations of eukaryotes, which preferentially consume and incorporate zinc relative to prokaryotes, approximately a billion years after their origin (at the latest).

Diversity

Eukaryotes are organisms that range from microscopic single cells, such as picozoans under 3 micrometres across, to animals like the blue whale, weighing up to 190 tonnes and measuring up to 33.6 metres (110 ft) long, or plants like the coast redwood, up to 120 metres (390 ft) tall.

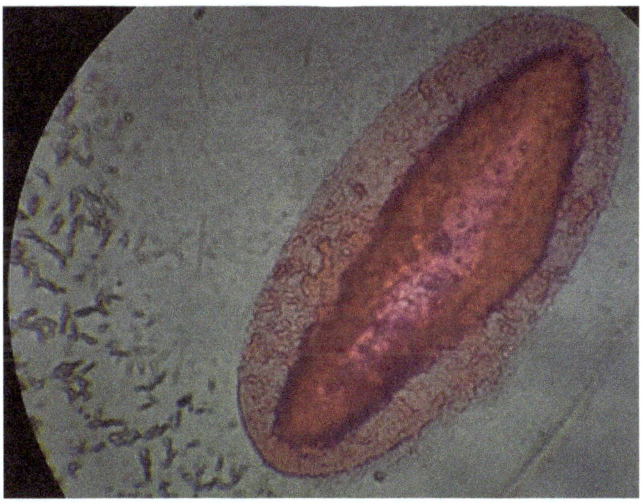

Fig 8: Prokaryotes (Small Cylindrical Cells, Bacteria, on Left) and a Single-Celled Eukaryote, *Paramecium*

Many eukaryotes are unicellular; the informal grouping called protists includes many of these, with some multicellular forms like the giant kelp up to 200 feet (61 m) long.

The multicellular eukaryotes include the animals, plants, and fungi, but again, these groups too contain many unicellular species. Eukaryotic cells are typically much larger than those of prokaryotes—the bacteria and the archaea—having a volume of around 10,000 times greater. Eukaryotes represent a small minority of the number of organisms, but, as many of them are much larger, their collective global biomass (468 gigatons) is far larger than that of prokaryotes (77 gigatons), with plants alone accounting for over 81% of the total biomass of Earth.

Fig 9: Coast Redwood

Fig 10: Blue Whale

The eukaryotes are a diverse lineage, consisting mainly of microscopic organisms. Multicellularity in some form has evolved independently at least 25 times within the eukaryotes. Complex multicellular organisms, not counting the aggregation of amoebae to form slime molds, have evolved within only six eukaryotic lineages: animals, symbiomycotan fungi, brown algae, red algae, green algae, and land plants. Eukaryotes are grouped by genomic similarities, so that groups often lack visible shared characteristics.

Distinguishing Features

Nucleus

The defining feature of eukaryotes is that their cells have nuclei. Eukaryotic cells have a variety of internal membrane-bound structures, called organelles, and a cytoskeleton which defines the cell's organization and shape. The nucleus stores the cell's DNA, which is divided into linear bundles called chromosomes; these are separated into two matching sets by a microtubular spindle during nuclear division, in the distinctively eukaryotic process of mitosis.

Biochemistry

Eukaryotes differ from prokaryotes in multiple ways, with unique biochemical pathways such as sterane synthesis. The eukaryotic signature proteins have no homology to proteins in other domains of life, but appear to be universal among eukaryotes. They include the proteins of the cytoskeleton, the complex transcription machinery, the membrane-sorting systems, the nuclear pore, and some enzymes in the biochemical pathways.

Internal Membranes

Eukaryote cells include a variety of membrane-bound structures, together forming the endomembrane system. Simple compartments, called vesicles and vacuoles, can form by budding off other membranes. Many cells ingest food and other materials through a process of endocytosis, where the outer membrane invaginates and then pinches off to form a vesicle. Some cell products can leave in a vesicle through exocytosis.

The nucleus is surrounded by a double membrane known as the nuclear envelope, with nuclear pores that allow material to move in and out. Various tube- and sheet-like extensions of the nuclear membrane form the endoplasmic reticulum, which is involved in protein transport and maturation.

It includes the rough endoplasmic reticulum, covered in ribosomes which synthesize proteins; these enter the interior space or lumen. Subsequently, they generally enter vesicles, which bud off from the smooth endoplasmic reticulum. In most eukaryotes, these protein-carrying vesicles are released and further modified in stacks of flattened vesicles (cisternae), the Golgi apparatus. Vesicles may be specialized; for instance, lysosomes contain digestive enzymes that break down biomolecules in the cytoplasm.

Mitochondria

Mitochondria are organelles in eukaryotic cells. The mitochondrion is commonly called "the powerhouse of the cell", for its function providing energy by oxidising sugars or fats to produce the energy-storing molecule ATP. Mitochondria have two surrounding membranes, each a phospholipid bilayer, the inner of which is folded into invaginations called cristae where aerobic respiration takes place.

Mitochondria contain their own DNA, which has close structural similarities to bacterial DNA, from which it originated, and which encodes rRNA and tRNA genes that produce RNA which is closer in structure to bacterial RNA than to eukaryote RNA.

Some eukaryotes, such as the metamonads *Giardia* and *Trichomonas*, and the amoebozoan *Pelomyxa*, appear to lack mitochondria, but all contain mitochondrion-derived organelles, like hydrogenosomes or mitosomes, having lost their mitochondria secondarily. They obtain energy by enzymatic action in the cytoplasm.

Plastids

Plants and various groups of algae have plastids as well as mitochondria. Plastids, like mitochondria, have their own DNA and are developed from endosymbionts, in this case cyanobacteria. They usually take the form of chloroplasts which, like cyanobacteria, contain chlorophyll and produce organic compounds (such as glucose) through photosynthesis. Others are involved in storing food. Although plastids probably had a single origin, not all plastid-containing groups are closely related. Instead, some eukaryotes have obtained them from others through secondary endosymbiosis or ingestion. The capture and sequestering of photosynthetic cells and chloroplasts, kleptoplasty, occurs in many types of modern eukaryotic organisms.

Cytoskeletal Structures

The cytoskeleton provides stiffening structure and points of attachment for motor structures that enable the cell to move, change shape, or transport materials. The motor structures are microfilaments of actin and actin-binding proteins, including α-actinin, fimbrin, and filamin are present in submembranous cortical layers and bundles. Motor proteins of microtubules, dynein and kinesin, and myosin of actin filaments, provide dynamic character of the network.

Many eukaryotes have long slender motile cytoplasmic projections, called flagella, or multiple shorter structures called cilia. These organelles are variously involved in movement, feeding, and sensation. They are composed mainly of tubulin, and are entirely distinct from prokaryotic flagella. They are supported by a bundle of microtubules arising from a centriole, characteristically arranged as nine doublets surrounding two singlets. Flagella may have hairs (mastigonemes), as in many Stramenopiles. Their interior is continuous with the cell's cytoplasm.

Centrioles are often present, even in cells and groups that do not have flagella, but conifers and flowering plants have neither. They generally occur in groups that give rise to various microtubular roots. These form a primary component of the cytoskeleton, and are often assembled over the course of several cell divisions, with one flagellum retained from the parent and the other derived from it. Centrioles produce the spindle during nuclear division.

Cell Wall

The cells of plants, algae, fungi and most chromalveolates, but not animals, are surrounded by a cell wall. This is a layer outside the cell membrane, providing the cell with structural support, protection, and a filtering mechanism. The cell wall also prevents over-expansion when water enters the cell.

The major polysaccharides making up the primary cell wall of land plants are cellulose, hemicellulose, and pectin. The cellulose microfibrils are linked together with hemicellulose, embedded in a pectin matrix. The most common hemicellulose in the primary cell wall is xyloglucan.

Sexual Reproduction

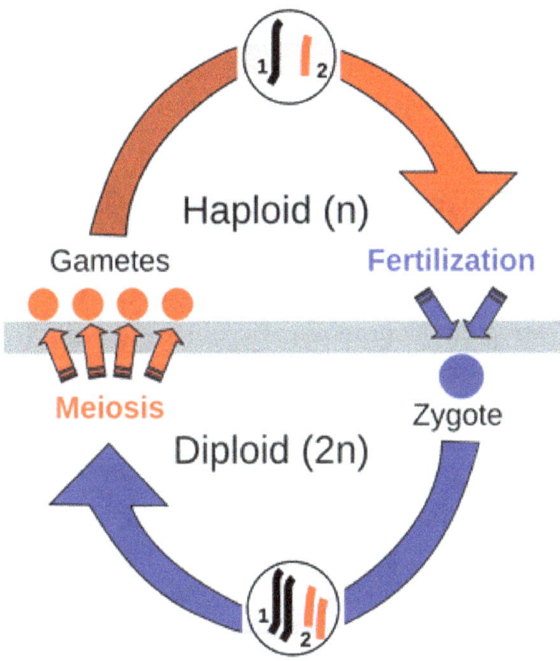

Fig 11: Sexual Reproduction Requires a Life Cycle that Alternates between a Haploid Phase, with One Copy of Each Chromosome in the Cell, and a Diploid Phase, with Two Copies. In Eukaryotes, Haploid Gametes are Produced by Meiosis; Two Gametes Fuse to Form a Diploid Zygote

Eukaryotes have a life cycle that involves sexual reproduction, alternating between a haploid phase, where only one copy of each chromosome is present in each cell, and a diploid phase, with two copies of each chromosome in each cell. The diploid phase is formed by fusion of two haploid gametes, such as eggs and spermatozoa, to form a zygote; this may grow into a body, with its cells dividing by mitosis, and at some stage produce haploid gametes through meiosis, a division that reduces the number of chromosomes and creates genetic variability. There is considerable variation in this pattern.

Plants have both haploid and diploid multicellular phases. Eukaryotes have lower metabolic rates and longer generation times than prokaryotes, because they are larger and therefore have a smaller surface area to volume ratio.

The evolution of sexual reproduction may be a primordial characteristic of eukaryotes. Based on a phylogenetic analysis, Dacks and Roger have proposed that facultative sex was present in the group's common ancestor. A core set of genes that function in meiosis is present in both *Trichomonas vaginalis* and *Giardia intestinalis*, two organisms previously thought to be asexual. Since these two species are descendants of lineages that diverged early from the eukaryotic evolutionary tree, core meiotic genes, and hence sex, were likely present in the common ancestor of eukaryotes.

Species once thought to be asexual, such as *Leishmania* parasites, have a sexual cycle. Amoebae, previously regarded as asexual, may be anciently sexual; while present-day asexual groups could have arisen recently.

2.3 MICROBIAL GROWTH AND REPRODUCTION: MECHANISMS OF MULTIPLICATION

Microorganisms exhibit unique growth and reproductive mechanisms, enabling them to colonize diverse environments. Microbial growth refers to an increase in cell number, and this process can occur through various methods, depending on the type of microorganism.

2.3.1 Binary Fission

Binary fission ("division in half") is a kind of asexual reproduction. It is the most common form of reproduction in prokaryotes such as bacteria. It occurs in some single-celled Eukaryotes like the *Amoeba* and the *Paramoecium*. In binary fission DNA replication and segregation occur simultaneously. In binary fission, the fully grown parent cell splits into two halves, producing two pools. After replicating its genetic material, the parent cell divides into two equal sized daughter cells. The genetic material is replicated, then equally split. The daughter cells are genetically identical (unless a mutation occurs during replication).

During binary fission, the DNA molecule divides and forms two DNA molecules. The cell then grows to create room for each molecule to move towards the opposite side of the bacterium. At the same time, the cell membrane divides to form 2 daughter cells. After division, the new cells grow and the process repeats itself. Binary fission occurs in eukaryotic tissue cells, but the process is more complicated. Basically, a cell copies its DNA, and then splits down the middle, creating two daughter cells.

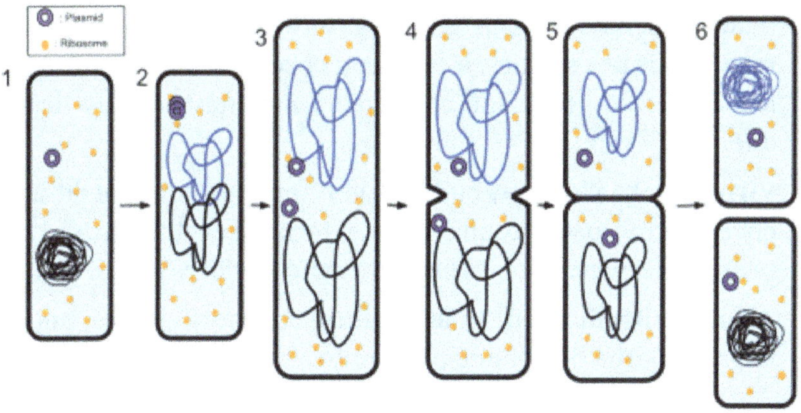

2.3.2 Budding

Budding or blastogenesis is a type of asexual reproduction in which a new organism develops from an outgrowth or bud due to cell division at one particular site. For example, the small bulb-like projection coming out from the yeast cell is known as a bud. Since the reproduction is asexual, the newly created organism is a clone and excepting mutations is genetically identical to the parent organism. Organisms such as hydra use regenerative cells for reproduction in the process of budding.

In hydra, a bud develops as an outgrowth due to repeated cell division at one specific site. These buds develop into tiny individuals and, when fully mature, detach from the parent body and become new independent individuals.

Internal budding or endodyogeny is a process of asexual reproduction, favored by parasites such as *Toxoplasma gondii*. It involves an unusual process in which two daughter cells are produced inside a mother cell, which is then consumed by the offspring prior to their separation. Endopolygeny is the division into several organisms at once by internal budding.

- **Cellular Reproduction:** Some cells divide asymmetrically by budding, for example *Saccharomyces cerevisiae*, the yeast species used in baking and brewing. This process results in a 'mother' cell and a smaller 'daughter' cell. Cryo-electron tomography recently revealed that mitochondria in cells divide by budding.
- **Animal Reproduction:** In some multicellular animals, offspring may develop as outgrowths of the mother. Animals that reproduce by budding include corals, some sponges, some acoels (e.g., *Convolutriloba*), echinoderm larvae, placozoans, symbions, pterobranchians,

entoproctans, some polychaetes, bryozoans, tunicates, flatworms and a single phoronid species.

- **Colony Division:** Colonies of some bee species have also exhibited budding behavior, such as *Apis dorsata*. Although budding behavior is rare in this bee species, it has been observed when a group of workers leave the natal nest and construct a new nest usually near the natal one.
- **Virology:** In virology, budding is a form of viral shedding by which enveloped viruses acquire their external envelope from the host cell membrane, which bulges outwards and encloses the virion.
- **Plant Multiplication:** In agriculture and horticulture, budding refers to grafting the bud of one plant onto another.

2.3.3 Sporulation in Bacillus Subtilis

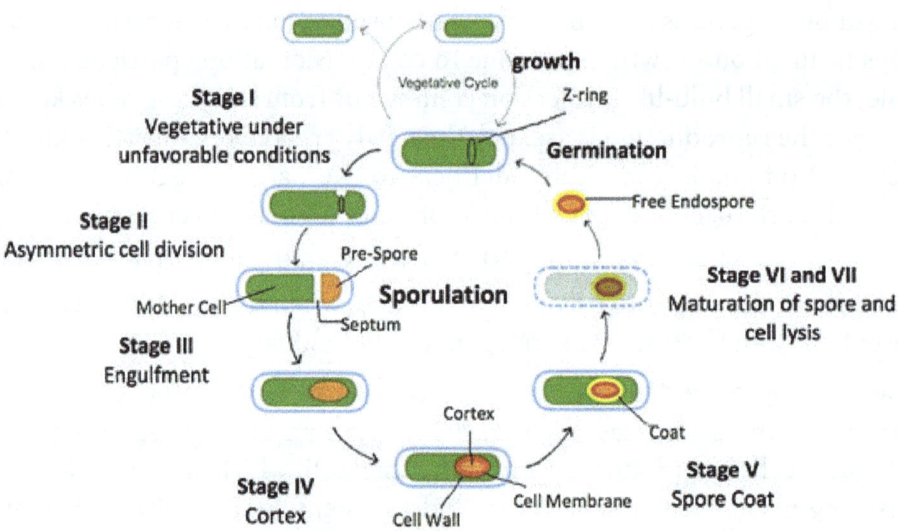

Fig 12: The Sporulation Process of Bacillus Subtilis

Bacillus subtilis is a rod-shaped, Gram-positive bacteria that is naturally found in soil and vegetation, and is known for its ability to form a small, tough, protective and metabolically dormant endospore.

B. subtilis can divide symmetrically to make two daughter cells (binary fission), or asymmetrically, producing a single endospore that is resistant to environmental factors such as heat, desiccation, radiation and chemical insult which can persist in the environment for long periods of time. The endospore is formed at times of nutritional stress, allowing the organism to persist in the environment until conditions become favourable. The process of endospore formation has profound morphological and physiological consequences: radical post-replicative remodelling of two progeny cells, accompanied eventually by cessation of metabolic activity in one daughter cell (the spore) and death by lysis of the other (the 'mother cell').

Commitment to Sporulation

Although sporulation in *B. subtilis* is induced by starvation, the sporulation developmental program is not initiated immediately when growth slows due to nutrient limitation.

A variety of alternative responses can occur, including the activation of flagellar motility to seek new food sources by chemotaxis, the production of antibiotics to destroy competing soil microbes, the secretion of hydrolytic enzymes to scavenge extracellular proteins and polysaccharides, or the induction of 'competence' for uptake of exogenous DNA for consumption, with the occasional side-effect that new genetic information is stably integrated.

Sporulation is the last-ditch response to starvation and is suppressed until alternative responses prove inadequate. Even then, certain conditions must be met such as chromosome integrity, the state of chromosomal replication, and the functioning of the Krebs cycle.

Nature of Regulation

Sporulation requires a great deal of time and also a lot of energy and is essentially irreversible, making it crucial for a cell to monitor its surroundings efficiently and ensure that sporulation is embarked upon at only the most appropriate times.

The wrong decision can be catastrophic: a vegetative cell will die if the conditions are too harsh, while bacteria forming spores in an environment which is conducive to vegetative growth will be out competed. In short, initiation of sporulation is a very tightly regulated network with numerous checkpoints for efficient control.

Control Points

Two transcriptional regulators, σH and Spo0A, play key roles in initiation of sporulation. Several additional proteins participate, mainly by controlling the accumulated concentration of Spo0A~P. Spo0A lies at the end of a series of inter-protein phosphotransfer reactions, Kin–Spo0F–Spo0B–Spo0A, termed as a 'phosphorelay'.

The term 'Activators' refers to genes/proteins that ultimately result in initiation of sporulation and 'Repressors' refers to the ones that inhibit this initiation of sporulation.

Activators	Regulation	Repressors	Regulation
KinA	Transfers phosphate to the Spo0F	Sda	Blocks autophosphorylation of KinA
KinB	Transfers phosphate to the Spo0F	KipI	Blocks autophosphorylation of KinA
Spo0A	Activates several key sporulation-specific genes	Spo0A	Negatively controls transcription of abrB
Spo0H (σH)	Activates phrE gene	ComA	Activates competence
Spo0B	Phosphotransferase initiation	SinR	Negative regulation of kinB
Spo0F	Phosphotransferase initiation	RapA, RapB, RapE, and RapH	Dephosphorylation of Spo0F~P
ComA	Activates phrA	Spo0E, YisI, and YnzD	Dephosphorylation of Spo0A~P
SinI	Antagonist of SinR	GTP-bound CodY	Inhibits rapA-phrA
KipA	Inhibits kipI gene	AbrB	Encodes a repressor of spo0H
PhrA	Suppresses dephosphorylation activity of RapA	Hpr	Overexpression inhibits sporulation
PhrE	Inhibits RapE	DnaA	Over expression of Sda
PhrH	Inhibits RapH		

Homologous Networks in Eukaryotes

Spores form a part of the life cycles of a diverse range of organisms such as many bacteria, plants, algae, fungi and some protozoa. In *Saccharomyces cerevisiae* (Kingdom Fungi), the set of early genes activating sporulation is induced by Ime1 (Inducer of Meiosis 1) and a regulator of middle genes is Ndt80p.

Similarly, in *Dictyostelium discoideum* (slime mold), SDF-2 and cytokinins are secreted during late development to trigger signal transduction pathways that lead to an increase in the activity of the cAMP-dependent protein kinase, PKA, which triggers rapid encapsulation as well as ensuring spore dormancy.

2.4 MICROBIAL METABOLISM: ENERGY PRODUCTION AND NUTRIENT UTILIZATION

Metabolism is the set of chemical reactions that occur within a microorganism to maintain life, providing energy for cellular functions and enabling growth and reproduction.

Microbial metabolism can vary widely, depending on the type of microorganism and its environmental conditions.

2.4.1 Catabolism

Catabolism is the set of metabolic pathways that breaks down molecules into smaller units that are either oxidized to release energy or used in other anabolic reactions. Catabolism breaks down large molecules (such as polysaccharides, lipids, nucleic acids, and proteins) into smaller units (such as monosaccharides, fatty acids, nucleotides, and amino acids, respectively). Catabolism is the breaking-down aspect of metabolism, whereas anabolism is the building-up aspect.

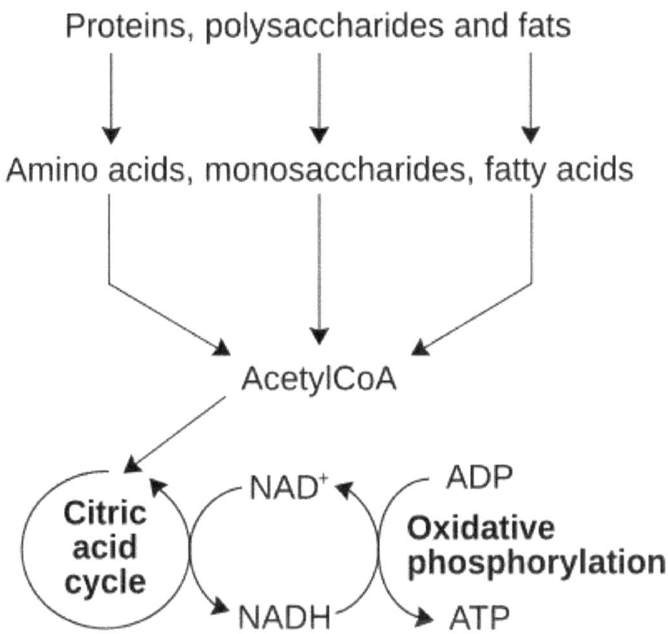

Fig 13: Schematic Breakdown of Large Biomolecules to Release Energy for Fueling the Cell Metabolism by Producing ATP, the Energy Currency of the Cell

Cells use the monomers released from breaking down polymers to either construct new polymer molecules or degrade the monomers further to simple waste products, releasing energy. Cellular wastes include lactic acid, acetic acid, carbon dioxide, ammonia, and urea.

The formation of these wastes is usually an oxidation process involving a release of chemical free energy, some of which is lost as heat, but the rest of which is used to drive the synthesis of adenosine triphosphate (ATP).

This molecule acts as a way for the cell to transfer the energy released by catabolism to the energy-requiring reactions that make up anabolism.

Catabolism is a destructive metabolism and anabolism is a constructive metabolism. Catabolism, therefore, provides the chemical energy necessary for the maintenance and growth of cells. Examples of catabolic processes include glycolysis, the citric acid cycle, the breakdown of muscle protein in order to use amino acids as substrates for gluconeogenesis, the breakdown of fat in adipose tissue to fatty acids, and oxidative deamination of neurotransmitters by monoamine oxidase.

Catabolic Hormones

There are many signals that control catabolism. Most of the known signals are hormones and the molecules involved in metabolism itself. Endocrinologists have traditionally classified many of the hormones as anabolic or catabolic, depending on which part of metabolism they stimulate.

The so-called classic catabolic hormones known since the early 20th century are cortisol, glucagon, and adrenaline (and other catecholamines).

In recent decades, many more hormones with at least some catabolic effects have been discovered, including cytokines, orexin (known as hypocretin), and melatonin.

Hormone	Function
Cortisol	Released from the adrenal gland in response to stress; its main role is to increase blood glucose levels by gluconeogenesis.
Glucagon	Released from alpha cells in the pancreas either when starving or when the body needs to generate additional energy; it stimulates the breakdown of glycogen in the liver to increase blood glucose levels; its effect is the opposite of insulin; glucagon and insulin are a part of a negative-feedback system that stabilizes blood glucose levels.
Adrenaline	Released in response to the activation of the sympathetic nervous system; increases heart rate and heart contractility, constricts blood vessels, is a bronchodilator that opens (dilates) the bronchi of the lungs to increase air volume and oxygen supply in the lungs, and stimulates gluconeogenesis.

2.4.2 Anabolism

Anabolism is the set of metabolic pathways that construct macromolecules like DNA or RNA from smaller units. These reactions require energy, known also as an endergonic process.

Anabolism is the building-up aspect of metabolism, whereas catabolism is the breaking-down aspect. Anabolism is usually synonymous with biosynthesis.

Pathway

Polymerization, an anabolic pathway used to build macromolecules such as nucleic acids, proteins, and polysaccharides, uses condensation reactions to join monomers. Macromolecules are created from smaller molecules using enzymes and cofactors.

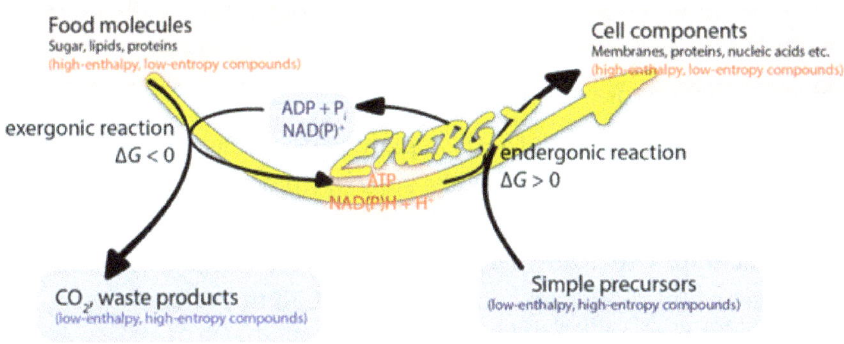

Fig 14: Use of ATP to Drive the Endergonic Process of Anabolism

Energy Source

Anabolism is powered by catabolism, where large molecules are broken down into smaller parts and then used up in cellular respiration. Many anabolic processes are powered by the cleavage of adenosine triphosphate (ATP). Anabolism usually involves reduction and decreases entropy, making it unfavorable without energy input. The starting materials, called the precursor molecules, are joined using the chemical energy made available from hydrolyzing ATP, reducing the cofactors NAD^+, $NADP^+$, and FAD, or performing other favorable side reactions. Occasionally it can also be driven by entropy without energy input, in cases like the formation of the phospholipid bilayer of a cell, where hydrophobic interactions aggregate the molecules.

Cofactors

The reducing agents NADH, NADPH, and $FADH_2$, as well as metal ions, act as cofactors at various steps in anabolic pathways. NADH, NADPH, and $FADH_2$ act as electron carriers, while charged metal ions within enzymes stabilize charged functional groups on substrates.

Substrates

Substrates for anabolism are mostly intermediates taken from catabolic pathways during periods of high energy charge in the cell.

Functions

Anabolic processes build organs and tissues. These processes produce growth and differentiation of cells and increase in body size, a process that involves synthesis of complex molecules. Examples of anabolic processes include the growth and mineralization of bone and increases in muscle mass.

Anabolic Hormones

Endocrinologists have traditionally classified hormones as anabolic or catabolic, depending on which part of metabolism they stimulate. The classic anabolic hormones are the anabolic steroids, which stimulate protein synthesis and muscle growth, and insulin.

Photosynthetic Carbohydrate Synthesis

Photosynthetic carbohydrate synthesis in plants and certain bacteria is an anabolic process that produces glucose, cellulose, starch, lipids, and proteins from CO_2. It uses the energy produced from the light-driven reactions of photosynthesis, and creates the precursors to these large molecules via carbon assimilation in the photosynthetic carbon reduction cycle, a.k.a. the Calvin cycle.

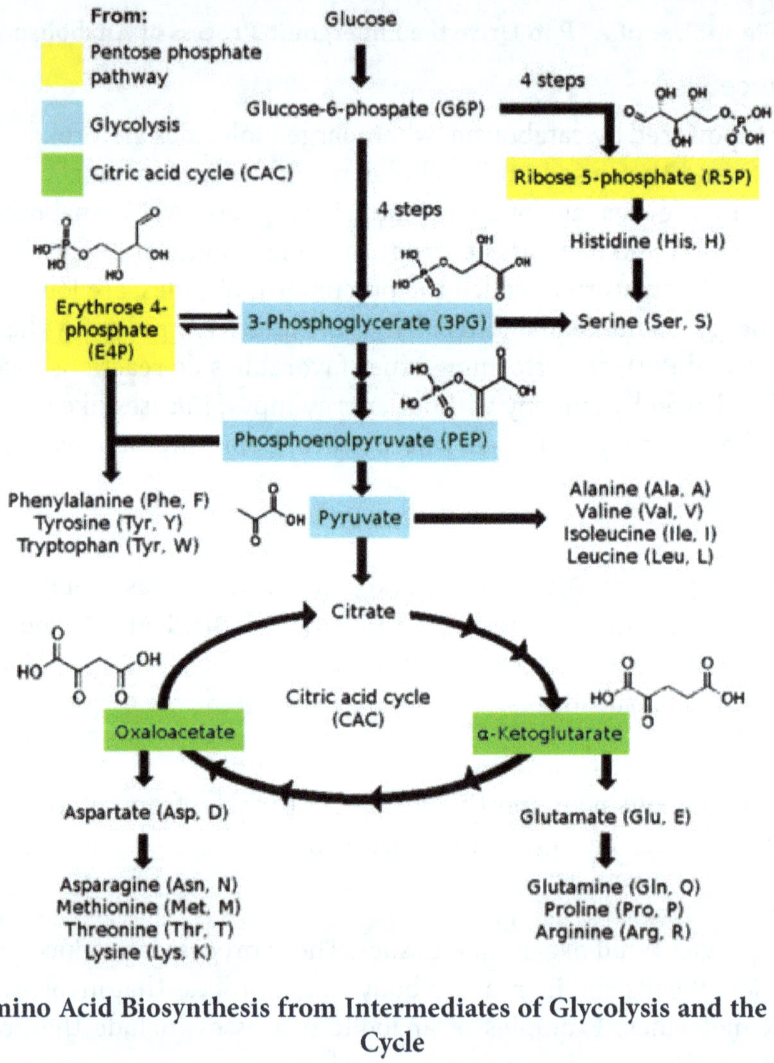

Fig 15: Amino Acid Biosynthesis from Intermediates of Glycolysis and the Citric Acid Cycle

Amino Acid Biosynthesis

All amino acids are formed from intermediates in the catabolic processes of glycolysis, the citric acid cycle, or the pentose phosphate pathway. From glycolysis, glucose

6-phosphate is a precursor for histidine; 3-phosphoglycerate is a precursor for glycine and cysteine; phosphoenol pyruvate, combined with the *3-phosphoglycerate*-derivative erythrose 4-phosphate, forms tryptophan, phenylalanine, and tyrosine; and pyruvate is a precursor for alanine, valine, leucine, and isoleucine. From the citric acid cycle, α-ketoglutarate is converted into glutamate and subsequently glutamine, proline, and arginine; and oxaloacetate is converted into aspartate and subsequently asparagine, methionine, threonine, and lysine.

Glycogen Storage

During periods of high blood sugar, glucose 6-phosphate from glycolysis is diverted to the glycogen-storing pathway. It is changed to glucose-1-phosphate by phosphoglucomutase and then to UDP-glucose by UTP--glucose-1-phosphate uridylyltransferase. Glycogen synthase adds this UDP-glucose to a glycogen chain.

Gluconeogenesis

Glucagon is traditionally a catabolic hormone, but also stimulates the anabolic process of gluconeogenesis by the liver, and to a lesser extent the kidney cortex and intestines, during starvation to prevent low blood sugar. It is the process of converting pyruvate into glucose. Pyruvate can come from the breakdown of glucose, lactate, amino acids, or glycerol. The gluconeogenesis pathway has many reversible enzymatic processes in common with glycolysis, but it is not the process of glycolysis in reverse. It uses different irreversible enzymes to ensure the overall pathway runs in one direction only.

Regulation

Anabolism operates with separate enzymes from catalysis, which undergo irreversible steps at some point in their pathways. This allows the cell to regulate the rate of production and prevent an infinite loop, also known as a futile cycle, from forming with catabolism.

The balance between anabolism and catabolism is sensitive to ADP and ATP, otherwise known as the energy charge of the cell. High amounts of ATP cause cells to favor the anabolic pathway and slow catabolic activity, while excess ADP slows anabolism and favors catabolism. These pathways are also regulated by circadian rhythms, with processes such as glycolysis fluctuating to match an animal's normal periods of activity throughout the day.

2.4.3 Nutritional Diversity

Nutritional diversity refers to the variety of nutrients that organisms require to function optimally and the broad range of dietary sources that provide these nutrients. A diet rich in diverse nutrients is fundamental to maintaining health, preventing chronic diseases, and promoting longevity. Nutritional diversity spans the essential macronutrients and micronutrients, phytochemicals, and fiber that come from varied foods across different food groups.

Nutritional diversity is a fundamental aspect of a healthy diet and plays a significant role in promoting long-term health, preventing chronic diseases, and supporting overall well-being.

By consuming a variety of foods across all food groups, individuals can ensure they are meeting their nutritional needs, from macronutrients and micronutrients to essential bioactive compounds. Despite challenges in accessibility and awareness, efforts to promote dietary diversity are critical in improving public health. Through education, policy support, and individual dietary choices, achieving nutritional diversity is possible, leading to healthier individuals and communities.

Importance of Nutritional Diversity

Nutritional diversity is critical because no single food can provide all the nutrients required for optimal health. A diverse diet helps ensure a balanced intake of carbohydrates, proteins, fats, vitamins, minerals, and other vital compounds necessary for growth, energy, and cellular repair.

Eating a variety of foods provides the body with antioxidants, phytochemicals, and bioactive compounds that work synergistically to protect against oxidative stress and inflammation, reducing the risk of chronic diseases.

- **Prevention of Nutrient Deficiency**: Nutrient deficiencies, such as iron deficiency anemia, vitamin A deficiency, and iodine deficiency, are common health problems resulting from a lack of dietary diversity. Consuming a variety of foods helps to prevent such deficiencies and promotes healthy growth and development, especially in children and pregnant women.
- **Support for Immune Function**: Nutritionally diverse diets are known to strengthen the immune system by providing necessary vitamins (such as vitamins A, C, and D) and minerals (like zinc and selenium), which play direct roles in immune function. A well-supported immune system is better equipped to fight off infections and recover from illness.
- **Reduced Risk of Chronic Diseases**: Nutritional diversity contributes to the prevention of chronic diseases like obesity, type 2 diabetes, cardiovascular diseases, and certain cancers. A varied diet rich in whole grains, fruits, vegetables, lean proteins, and healthy fats can help maintain healthy blood pressure, cholesterol levels, and blood glucose levels, all of which are factors in chronic disease prevention.

Components of Nutritional Diversity

Nutritional diversity consists of various food groups that provide unique combinations of essential nutrients. Each macronutrient, micronutrient, and bioactive compound has distinct functions that collectively support physiological processes.

- **Macronutrients:** These include carbohydrates, proteins, and fats, which provide energy and structural components for the body. Carbohydrates offer a quick source of energy, proteins are vital for tissue repair and immune function, and fats are crucial for brain health and hormone production. Including a variety of grains, legumes, nuts, seeds, and oils in the diet ensures a balanced intake of these macronutrients.
- **Micronutrients:** Vitamins and minerals are necessary in smaller amounts but are vital for metabolic functions, bone health, and cellular repair. For example, vitamin C supports skin health and immune defense, while calcium and vitamin D are essential for strong bones. Eating a range of fruits, vegetables, dairy products, and fortified foods can help meet these requirements.
- **Phytochemicals and Antioxidants:** These are naturally occurring compounds in plant-based foods, including fruits, vegetables, nuts, and whole grains. Phytochemicals like flavonoids, carotenoids, and polyphenols have antioxidant properties that protect cells from damage, reduce inflammation, and may prevent diseases like cancer. Diverse plant foods are the primary source of these compounds.
- **Fiber:** Dietary fiber, found in plant foods such as whole grains, vegetables, fruits, and legumes, plays a critical role in digestive health. Soluble fiber helps regulate blood glucose and cholesterol, while insoluble fiber promotes regular bowel movements. A varied intake of fiber-rich foods supports gut health and contributes to overall well-being.

Sources of Nutritional Diversity

Achieving nutritional diversity requires consuming foods from all food groups, each providing unique nutrients that contribute to balanced health.

- **Fruits and Vegetables:** These are rich in vitamins, minerals, fiber, and phytochemicals. The color variety in fruits and vegetables (red, orange, green, purple) often indicates different beneficial compounds, such as beta-carotene in carrots and anthocyanins in blueberries. A colorful diet ensures a mix of nutrients that support various health functions.

- **Whole Grains and Legumes**: Whole grains, like brown rice, quinoa, and oats, and legumes, such as beans and lentils, provide complex carbohydrates, fiber, protein, and essential minerals. They contribute to sustained energy, gut health, and cardiovascular health. Including a range of grains and legumes ensures a comprehensive intake of nutrients.
- **Protein Sources**: Both animal (meat, fish, eggs, dairy) and plant-based (soy, lentils, nuts) proteins are essential for muscle development, immune health, and enzyme production. A diverse protein intake helps meet amino acid requirements while offering other nutrients, such as iron, calcium, and B vitamins, found in specific protein sources.
- **Healthy Fats**: Nutritionally diverse fats from sources like avocados, olive oil, nuts, and fatty fish (e.g., salmon) provide essential fatty acids, including omega-3 and omega-6. These fats support brain function, reduce inflammation, and improve heart health. Consuming different types of fats ensures a balance of essential fatty acids.

Challenges in Achieving Nutritional Diversity

Despite the known benefits, achieving nutritional diversity can be challenging due to economic, cultural, and lifestyle factors.

- **Economic Constraints**: Healthy and diverse foods, such as fresh fruits, vegetables, and lean proteins, may be less accessible or affordable for low-income populations, leading to dietary imbalances. Processed foods, which are often cheaper, lack the diversity of nutrients necessary for optimal health, contributing to nutritional deficiencies and obesity.
- **Food Deserts**: In certain urban and rural areas, access to a variety of fresh, healthy foods is limited. This lack of availability can hinder efforts to maintain a nutritionally diverse diet and increase dependence on packaged and processed foods.
- **Lack of Awareness**: Many individuals may not be aware of the importance of nutritional diversity or lack the knowledge of how to incorporate a variety of foods into their daily diet. Public health education plays a role in raising awareness about the benefits of diverse food choices.
- **Cultural and Dietary Preferences**: Personal, cultural, and religious preferences can influence dietary patterns and may restrict the range of foods consumed. While these preferences are important, finding

ways to diversify within dietary restrictions is essential for achieving nutritional balance.

Promoting Nutritional Diversity: Strategies and Recommendations

Efforts to promote nutritional diversity involve education, policy, and individual choices. Implementing these strategies can improve health outcomes and increase the quality of diets worldwide.

- **Public Health Initiatives:** Governments and health organizations can encourage nutritional diversity through policies, such as subsidizing fruits and vegetables, and implementing school nutrition programs. Public health campaigns can also raise awareness about the importance of a diverse diet and provide practical tips on incorporating a variety of foods.
- **Dietary Education:** Educating individuals on how to build nutritionally diverse meals is crucial. For instance, the "plate method" encourages individuals to fill half of their plate with fruits and vegetables, a quarter with whole grains, and the remaining quarter with protein. These guidelines help individuals make balanced food choices.
- **Encouraging Local and Seasonal Eating:** Consuming locally grown and seasonal foods supports nutritional diversity by introducing a range of fresh, nutrient-rich foods throughout the year. Seasonal diets often include a natural variety of nutrients, as different crops become available at different times.
- **Incorporating Global Food Varieties:** Exploring cuisines from different cultures can enhance dietary diversity and introduce new nutrient sources. For example, traditional Mediterranean, Asian, and African cuisines emphasize varied fruits, vegetables, whole grains, and legumes, offering a model for a balanced, nutritionally rich diet.

2.5 MICROBIAL GENETICS AND EVOLUTION: ADAPTATION AND DIVERSITY

Microbial genetics is a fundamental area of microbiology that studies the genetic makeup, gene expression, and evolution of microorganisms. Due to their rapid reproduction and high mutation rates, microorganisms can adapt quickly to changing environments, which contributes to their genetic diversity.

Microbes can exchange genetic material through several mechanisms, including:

- **Transformation:** The uptake of free DNA from the environment.
- **Conjugation:** The transfer of DNA between bacterial cells through a direct connection.

- **Transduction:** The transfer of genetic material by viruses, known as bacteriophages.

These genetic exchanges contribute to the development of antibiotic resistance in pathogenic bacteria, posing challenges for public health. Understanding microbial genetics helps researchers develop new treatments and strategies to combat resistant strains, as well as harness beneficial microorganisms for biotechnological applications.

2.6 INTERACTIONS OF MICROORGANISMS WITH THE ENVIRONMENT

Microorganisms interact with their environment in various ways, playing essential roles in nutrient cycles, decomposition, and ecosystem health. They contribute to carbon, nitrogen, and sulfur cycles by breaking down organic matter and releasing nutrients into the soil, supporting plant growth.

In aquatic environments, microorganisms form the base of the food chain, sustaining various marine organisms. Microbial symbiosis is also seen in plant and animal relationships, where both organisms benefit from the interaction. For example, nitrogen-fixing bacteria in the root nodules of legumes convert atmospheric nitrogen into forms usable by plants, enriching soil fertility.

Microbial interactions extend to human health, as the human body hosts diverse microorganisms collectively known as the microbiome. These microorganisms aid in digestion, immune function, and protection against pathogens. Disruption of the microbiome, often due to antibiotic use or dietary changes, can lead to health issues, highlighting the importance of balanced microbial communities.

2.7 APPLIED MICROBIOLOGY: INDUSTRIAL, MEDICAL, AND ENVIRONMENTAL APPLICATIONS

Microbiology has numerous applications in industry, medicine, and environmental management. In the medical field, microbiologists develop vaccines, antibiotics, and diagnostic techniques to detect and combat diseases. In agriculture, beneficial microorganisms are used to improve soil fertility, enhance crop protection, and control pests.

Industrial microbiology involves the use of microbes in processes such as fermentation, which is used to produce food items like yogurt, cheese, and alcoholic beverages. Environmental microbiology focuses on bioremediation, where microbes are used to break down pollutants and treat wastewater, helping to reduce environmental damage.

2.8 CONCLUSION

Microbiology is a broad and dynamic field that encompasses the study of microorganisms, their structures, functions, and roles in various ecosystems. From microbial diversity and growth to genetic exchange and environmental interactions, the fundamental features of microbiology provide insights into how microorganisms shape the natural world and impact human society.

Advances in microbiology continue to contribute to healthcare, agriculture, industry, and environmental sustainability, underscoring the importance of understanding and harnessing microbial life for the benefit of humanity.

Through continued research and innovation, microbiology will remain a key player in addressing global challenges and improving the quality of life on Earth.

CHAPTER 3
BACTERIA

Bacteria are microscopic, single-celled organisms that have a profound impact on all forms of life. Though often associated with disease, bacteria are incredibly diverse and play essential roles in a wide range of environments. Found virtually everywhere—from the soil and water to extreme environments like hot springs and ice caps—bacteria are crucial for nutrient cycling, environmental stability, and even human health.

3.1 UNDERSTANDING BACTERIAL STRUCTURE AND COMPOSITION

Bacteria are prokaryotic organisms, meaning they lack a true nucleus and membrane-bound organelles. Instead, they have simpler cellular structures compared to eukaryotic cells, like those of plants and animals. A typical bacterial cell has a cell wall, plasma membrane, cytoplasm, and genetic material.

3.1.1 Cell Wall

A cell wall is a structural layer that surrounds some cell types, found immediately outside the cell membrane. It can be tough, flexible, and sometimes rigid. Primarily, it provides the cell with structural support, shape, protection, and functions as a selective barrier. Another vital role of the cell wall is to help the cell withstand osmotic pressure and mechanical stress. While absent in many eukaryotes, including animals, cell walls are prevalent in other organisms such as fungi, algae and plants, and are commonly found in most prokaryotes, with the exception of mollicute bacteria.

The composition of cell walls varies across taxonomic groups, species, cell type, and the cell cycle. In land plants, the primary cell wall comprises polysaccharides like cellulose, hemicelluloses, and pectin. Often, other polymers such as lignin, suberin or cutin are anchored to or embedded in plant cell walls. Algae exhibit cell walls composed of glycoproteins and polysaccharides, such as carrageenan and agar, distinct from those in land plants. Bacterial cell walls contain peptidoglycan, while archaeal cell walls vary in composition, potentially consisting of glycoprotein S-layers, pseudopeptidoglycan, or polysaccharides. Fungi possess cell walls constructed from the polymer chitin, specifically N-acetylglucosamine. diatoms have a unique cell wall composed of biogenic silica.

History

A plant cell wall was first observed and named (simply as a "wall") by Robert Hooke in 1665. However, "the dead excrusion product of the living protoplast" was forgotten, for almost three centuries, being the subject of scientific interest mainly as a resource for industrial processing or in relation to animal or human health.

In 1804, Karl Rudolphi and J.H.F. Link proved that cells had independent cell walls. Before, it had been thought that cells shared walls and that fluid passed between them this way.

The mode of formation of the cell wall was controversial in the 19th century. Hugo von Mohl (1853, 1858) advocated the idea that the cell wall grows by apposition. Carl Nägeli (1858, 1862, 1863) believed that the growth of the wall in thickness and in area was due to a process termed intussusception. Each theory was improved in the following decades: the apposition (or lamination) theory by Eduard Strasburger (1882, 1889), and the intussusception theory by Julius Wiesner (1886).

In 1930, Ernst Münch coined the term *apoplast* in order to separate the "living" symplast from the "dead" plant region, the latter of which included the cell wall.

By the 1980s, some authors suggested replacing the term "cell wall", particularly as it was used for plants, with the more precise term "extracellular matrix", as used for animal cells, but others preferred the older term.

Properties

Cell walls serve similar purposes in those organisms that possess them. They may give cells rigidity and strength, offering protection against mechanical stress. The chemical composition and mechanical properties of the cell wall are linked with plant cell growth and morphogenesis. In multicellular organisms, they permit the organism to build and hold a definite shape. Cell walls also limit the entry of large molecules that may be toxic to the cell. They further permit the creation of stable osmotic environments by preventing osmotic lysis and helping to retain water. Their composition, properties, and form may change during the cell cycle and depend on growth conditions.

Rigidity of Cell Walls

In most cells, the cell wall is flexible, meaning that it will bend rather than holding a fixed shape, but has considerable tensile strength. The apparent rigidity of primary plant tissues is enabled by cell walls, but is not due to the walls' stiffness. Hydraulic turgor pressure creates this rigidity, along with the wall structure. The flexibility of the cell walls is seen when plants wilt, so that the stems and leaves begin to droop, or in seaweeds that bend in water currents. As John Howland explains:

- Think of the cell wall as a wicker basket in which a balloon has been inflated so that it exerts pressure from the inside. Such a basket is very rigid and resistant to mechanical damage. Thus does the prokaryote cell (and eukaryotic cell that possesses a cell wall) gain strength from a flexible plasma membrane pressing against a rigid cell wall.

The apparent rigidity of the cell wall thus results from inflation of the cell contained within. This inflation is a result of the passive uptake of water. In plants, a secondary cell wall is a thicker additional layer of cellulose which increases wall rigidity. Additional layers may be formed by lignin in xylem cell walls, or suberin in cork cell walls. These compounds are rigid and waterproof, making the secondary wall stiff. Both wood and bark cells of trees have secondary walls. Other parts of plants such as the leaf stalk may acquire similar reinforcement to resist the strain of physical forces.

Permeability

The primary cell wall of most plant cells is freely permeable to small molecules including small proteins, with size exclusion estimated to be 30-60 kDa. The pH is an important factor governing the transport of molecules through cell walls.

Evolution

The photosynthetic eukaryotes (so-called plant and algae) is one group with cellulose cell walls, where the cell wall is closely related to the evolution of multicellularity, terrestrialization and vascularization. The CesA cellulose synthase evolved in *Cyanobacteria* and was part of Archaeplastida since endosymbiosis; secondary endosymbiosis events transferred it (with the arabinogalactan proteins) further into brown algae and oomycetes. Plants later evolved various genes from CesA, including the Csl (cellulose synthase-like) family of proteins and additional Ces proteins. Combined with the various glycosyltransferases (GT), they enable more complex chemical structures to be built.

Fungi use a chitin-glucan-protein cell wall. They share the 1,3-β-glucan synthesis pathway with plants, using homologous GT48 family 1,3-Beta-glucan synthases to perform the task, suggesting that such an enzyme is very ancient within the eukaryotes. Their glycoproteins are rich in mannose. The cell wall might have evolved to deter viral infections. Proteins embedded in cell walls are variable, contained in tandem repeats subject to homologous recombination. An alternative scenario is that fungi started with a chitin-based cell wall and later acquired the GT-48 enzymes for the 1,3-β-glucans via horizontal gene transfer. The pathway leading to 1,6-β-glucan synthesis is not sufficiently known in either case.

Plant Cell Walls

The walls of plant cells must have sufficient tensile strength to withstand internal osmotic pressures of several times atmospheric pressure that result from the difference in solute concentration between the cell interior and external solutions. Plant cell walls vary from 0.1 to several µm in thickness.

Layers

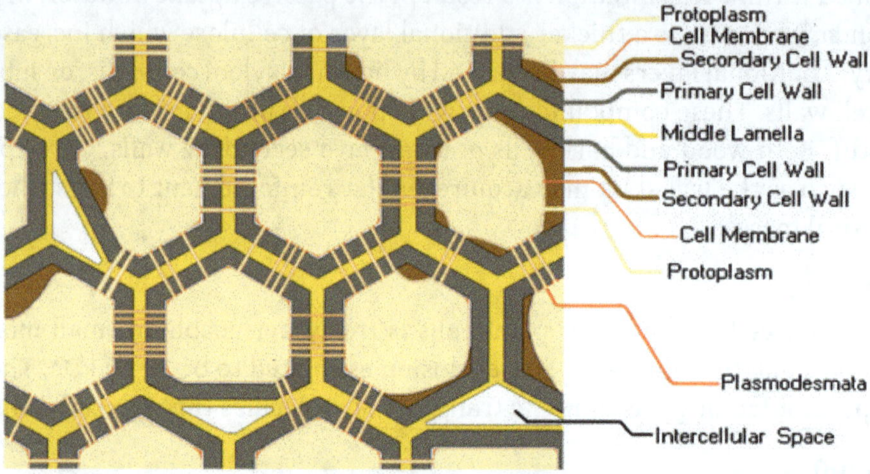

Fig 1: Cell Wall in Multicellular Plants—Its Different Layers and their Placement with Respect to Protoplasm

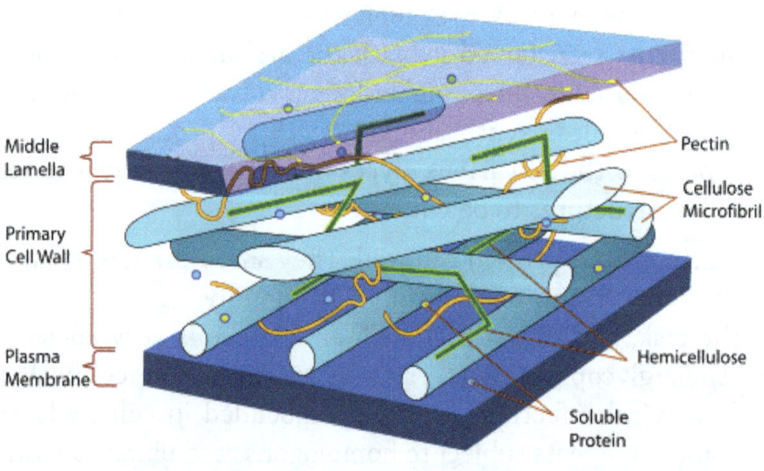

Fig 2: Molecular Structure of the Primary Cell Wall in Plants

Up to three strata or layers may be found in plant cell walls:

- **The primary cell wall**, generally a thin, flexible and extensible layer formed while the cell is growing.

- **The secondary cell wall**, a thick layer formed inside the primary cell wall after the cell is fully grown. It is not found in all cell types. Some cells, such as the conducting cells in xylem, possess a secondary wall containing lignin, which strengthens and waterproofs the wall.
- **The middle lamella**, a layer rich in pectins. This outermost layer forms the interface between adjacent plant cells and glues them together.

Composition

In the primary (growing) plant cell wall, the major carbohydrates are cellulose, hemicellulose and pectin. The cellulose microfibrils are linked via hemicellulosic tethers to form the cellulose-hemicellulose network, which is embedded in the pectin matrix. The most common hemicellulose in the primary cell wall is xyloglucan. In grass cell walls, xyloglucan and pectin are reduced in abundance and partially replaced by glucuronoarabinoxylan, another type of hemicellulose. Primary cell walls characteristically extend (grow) by a mechanism called acid growth, mediated by expansins, extracellular proteins activated by acidic conditions that modify the hydrogen bonds between pectin and cellulose. This functions to increase cell wall extensibility. The outer part of the primary cell wall of the plant epidermis is usually impregnated with cutin and wax, forming a permeability barrier known as the plant cuticle.

Secondary cell walls contain a wide range of additional compounds that modify their mechanical properties and permeability. The major polymers that make up wood (largely secondary cell walls) include:

- **Cellulose:** 35-50%.
- **Xylan:** 20-35%, a type of hemicellulose.
- **Lignin:** 10-25%, a complex phenolic polymer that penetrates the spaces in the cell wall between cellulose, hemicellulose and pectin components, driving out water and strengthening the wall.

Additionally, structural proteins (1-5%) are found in most plant cell walls; they are classified as hydroxyproline-rich glycoproteins (HRGP), arabinogalactan proteins (AGP), glycine-rich proteins (GRPs), and proline-rich proteins (PRPs). Each class of glycoprotein is defined by a characteristic, highly repetitive protein sequence. Most are glycosylated, contain hydroxyproline (Hyp) and become cross-linked in the cell wall.

These proteins are often concentrated in specialized cells and in cell corners. Cell walls of the epidermis may contain cutin. The Casparian strip in the endodermis roots and cork cells of plant bark contain suberin.

Both cutin and suberin are polyesters that function as permeability barriers to the movement of water. The relative composition of carbohydrates, secondary compounds and proteins varies between plants and between the cell type and age. Plant cells walls also contain numerous enzymes, such as hydrolases, esterases, peroxidases, and transglycosylases, that cut, trim and cross-link wall polymers.

Secondary walls - especially in grasses - may also contain microscopic silica crystals, which may strengthen the wall and protect it from herbivores.

Cell walls in some plant tissues also function as storage deposits for carbohydrates that can be broken down and resorbed to supply the metabolic and growth needs of the plant. For example, endosperm cell walls in the seeds of cereal grasses, nasturtium and other species, are rich in glucans and other polysaccharides that are readily digested by enzymes during seed germination to form simple sugars that nourish the growing embryo.

Formation

The middle lamella is laid down first, formed from the cell plate during cytokinesis, and the primary cell wall is then deposited inside the middle lamella. The actual structure of the cell wall is not clearly defined and several models exist - the covalently linked cross model, the tether model, the diffuse layer model and the stratified layer model. However, the primary cell wall, can be defined as composed of cellulose microfibrils aligned at all angles. Cellulose microfibrils are produced at the plasma membrane by the cellulose synthase complex, which is proposed to be made of a hexameric rosette that contains three cellulose synthase catalytic subunits for each of the six units. Microfibrils are held together by hydrogen bonds to provide a high tensile strength. The cells are held together and share the gelatinous membrane (the middle lamella), which contains magnesium and calcium pectates (salts of pectic acid). Cells interact though plasmodesmata, which are inter-connecting channels of cytoplasm that connect to the protoplasts of adjacent cells across the cell wall.

In some plants and cell types, after a maximum size or point in development has been reached, a *secondary wall* is constructed between the plasma membrane and primary wall. Unlike the primary wall, the cellulose microfibrils are aligned parallel in layers, the orientation changing slightly with each additional layer so that the structure becomes helicoidal. Cells with secondary cell walls can be rigid, as in the gritty sclereid cells in pear and quince fruit. Cell to cell communication is possible through pits in the secondary cell wall that allow plasmodesmata to connect cells through the secondary cell walls.

Fungal Cell Walls

There are several groups of organisms that have been called "fungi". Some of these groups (Oomycete and Myxogastria) have been transferred out of the Kingdom Fungi, in part because of fundamental biochemical differences in the composition of the cell wall. Most true fungi have a cell wall consisting largely of chitin and other polysaccharides. True fungi do not have cellulose in their cell walls.

In fungi, the cell wall is the outer-most layer, external to the plasma membrane. The fungal cell wall is a matrix of three main components:

- **Chitin:** Polymers consisting mainly of unbranched chains of β-(1,4)-linked-N-Acetylglucosamine in the Ascomycota and Basidiomycota, or poly-β-(1,4)-linked-N-Acetylglucosamine (chitosan) in the Zygomycota. Both chitin and chitosan are synthesized and extruded at the plasma membrane.
- **Glucans:** Glucose polymers that function to cross-link chitin or chitosan polymers. β-glucans are glucose molecules linked via β-(1,3)- or β-(1,6)-bonds and provide rigidity to the cell wall while α-glucans are defined by α-(1,3)- and/or α-(1,4) bonds and function as part of the matrix.
- **Proteins:** Enzymes necessary for cell wall synthesis and lysis in addition to structural proteins are all present in the cell wall. Most of the structural proteins found in the cell wall are glycosylated and contain mannose, thus these proteins are called mannoproteins or mannans.

Other Eukaryotic Cell Walls

Algae

Like plants, algae have cell walls. Algal cell walls contain either polysaccharides (such as cellulose (a glucan)) or a variety of glycoproteins (Volvocales) or both. The inclusion of additional polysaccharides in algal cells walls is used as a feature for algal taxonomy.

- **Mannans:** They form microfibrils in the cell walls of a number of marine green algae including those from the genera, *Codium*, *Dasycladus*, and *Acetabularia* as well as in the walls of some red algae, like *Porphyra* and *Bangia.*
- **Alginic Acid:** It is a common polysaccharide in the cell walls of brown algae.
- **Sulfonated Polysaccharides:** They occur in the cell walls of most algae; those common in red algae include agarose, carrageenan, porphyran, furcelleran and funoran.

Other compounds that may accumulate in algal cell walls include sporopollenin and calcium ions. The group of algae known as the diatoms synthesize their cell walls (also known as frustules or valves) from silicic acid. Significantly, relative to the organic cell walls produced by other groups, silica frustules require less energy to synthesize (approximately 8%), potentially a major saving on the overall cell energy budget and possibly an explanation for higher growth rates in diatoms. In brown algae, phlorotannins may be a constituent of the cell walls.

Water Molds

The group Oomycetes, also known as water molds, are saprotrophic plant pathogens like fungi. Until recently they were widely believed to be fungi, but structural and molecular evidence has led to their reclassification as heterokonts, related to autotrophic brown algae and diatoms. Unlike fungi, oomycetes typically possess cell walls of cellulose and glucans rather than chitin, although some genera (such as *Achlya* and *Saprolegnia*) do have chitin in their walls. The fraction of cellulose in the walls is no more than 4 to 20%, far less than the fraction of glucans. Oomycete cell walls also contain the amino acid hydroxyproline, which is not found in fungal cell walls.

Slime Molds

The dictyostelids are another group formerly classified among the fungi. They are slime molds that feed as unicellular amoebae, but aggregate into a reproductive stalk and sporangium under certain conditions. Cells of the reproductive stalk, as well as the spores formed at the apex, possess a cellulose wall. The spore wall has three layers, the middle one composed primarily of cellulose, while the innermost is sensitive to cellulase and pronase.

Prokaryotic Cell Walls

Bacterial Cell Walls

Around the outside of the cell membrane is the bacterial cell wall. Bacterial cell walls are made of peptidoglycan (also called murein), which is made from polysaccharide chains cross-linked by unusual peptides containing D-amino acids. Bacterial cell walls are different from the cell walls of plants and fungi which are made of cellulose and chitin, respectively. The cell wall of bacteria is also distinct from that of Archaea, which do not contain peptidoglycan. The cell wall is essential to the survival of many bacteria, although L-form bacteria can be produced in the laboratory that lack a cell wall.

The antibiotic penicillin is able to kill bacteria by preventing the cross-linking of peptidoglycan and this causes the cell wall to weaken and lyse. The lysozyme enzyme can also damage bacterial cell walls.

There are broadly speaking two different types of cell wall in bacteria, called gram-positive and gram-negative. The names originate from the reaction of cells to the Gram stain, a test long-employed for the classification of bacterial species.

Gram-positive bacteria possess a thick cell wall containing many layers of peptidoglycan and teichoic acids.

Gram-negative bacteria have a relatively thin cell wall consisting of a few layers of peptidoglycan surrounded by a second lipid membrane containing lipopolysaccharides and lipoproteins. Most bacteria have the gram-negative cell wall and only the Bacillota and Actinomycetota (previously known as the low G+C and high G+C gram-positive bacteria, respectively) have the alternative gram-positive arrangement.

These differences in structure produce differences in antibiotic susceptibility. The beta-lactam antibiotics (e.g., penicillin, cephalosporin) only work against gram-negative pathogens, such as *Haemophilus influenzae* or *Pseudomonas aeruginosa*. The glycopeptide antibiotics (e.g., vancomycin, teicoplanin, telavancin) only work against gram-positive pathogens such as *Staphylococcus aureus*

Archaeal Cell Walls

Although not truly unique, the cell walls of Archaea are unusual. Whereas peptidoglycan is a standard component of all bacterial cell walls, all archaeal cell walls lack peptidoglycan, though some methanogens have a cell wall made of a similar polymer called pseudopeptidoglycan. There are four types of cell wall currently known among the Archaea.

One type of archaeal cell wall is that composed of pseudopeptidoglycan (also called pseudomurein). This type of wall is found in some methanogens, such as *Methanobacterium* and *Methanothermus*. While the overall structure of archaeal *pseudo*peptidoglycan superficially resembles that of bacterial peptidoglycan, there are a number of significant chemical differences. Like the peptidoglycan found in bacterial cell walls, pseudopeptidoglycan consists of polymer chains of glycan cross-linked by short peptide connections. However, unlike peptidoglycan, the sugar N-acetylmuramic acid is replaced by N-acetyltalosaminuronic acid, and the two sugars are bonded with a β,1-3 glycosidic linkage instead of β,1-4. Additionally, the cross-linking peptides are L-amino acids rather than D-amino acids as they are in bacteria.

A second type of archaeal cell wall is found in *Methanosarcina* and *Halococcus*. This type of cell wall is composed entirely of a thick layer of polysaccharides, which may be sulfated in the case of *Halococcus*. Structure in this type of wall is complex and not fully investigated.

A third type of wall among the Archaea consists of glycoprotein, and occurs in the hyperthermophiles, *Halobacterium*, and some methanogens. In *Halobacterium*, the proteins in the wall have a high content of acidic amino acids, giving the wall an overall negative charge. The result is an unstable structure that is stabilized by the presence of large quantities of positive sodium ions that neutralize the charge. Consequently, *Halobacterium* thrives only under conditions with high salinity.

In other Archaea, such as *Methanomicrobium* and *Desulfurococcus*, the wall may be composed only of surface-layer proteins, known as an *S-layer*. S-layers are common in bacteria, where they serve as either the sole cell-wall component or an outer layer in conjunction with polysaccharides. Most Archaea are Gram-negative, though at least one Gram-positive member is known.

Other Cell Coverings

Many protists and bacteria produce other cell surface structures apart from cell walls, external (extracellular matrix) or internal. Many algae have a sheath or envelope of mucilage outside the cell made of exopolysaccharides. Diatoms build a frustule from silica extracted from the surrounding water; radiolarians, foraminiferans, testate amoebae and silicoflagellates also produce a skeleton from minerals, called test in some groups. Many green algae, such as *Halimeda* and the Dasycladales, and some red algae, the Corallinales, encase their cells in a secreted skeleton of calcium carbonate. In each case, the wall is rigid and essentially inorganic. It is the non-living component of cell. Some golden algae, ciliates and choanoflagellates produces a shell-like protective outer covering called lorica. Some dinoflagellates have a theca of cellulose plates, and coccolithophorids have coccoliths. An extracellular matrix (ECM) is also present in metazoans. Its composition varies between cells, but collagens are the most abundant protein in the ECM.

3.1.2 Cell Membrane

The cell membrane is a biological membrane that separates and protects the interior of a cell from the outside environment (the extracellular space). The cell membrane consists of a lipid bilayer, made up of two layers of phospholipids with cholesterols (a lipid component) interspersed between them, maintaining appropriate membrane fluidity at various temperatures. The membrane also contains membrane proteins, including integral proteins that span the membrane and serve as membrane transporters, and peripheral proteins that loosely attach to the outer (peripheral) side of the cell membrane, acting as enzymes to facilitate interaction with the cell's environment. Glycolipids embedded in the outer lipid layer serve a similar purpose.

The cell membrane controls the movement of substances in and out of a cell, being selectively permeable to ions and organic molecules. In addition, cell membranes are involved in a variety of cellular processes such as cell adhesion, ion conductivity, and cell signalling and serve as the attachment surface for several extracellular structures, including the cell wall and the carbohydrate layer called the glycocalyx, as well as the intracellular network of protein fibers called the cytoskeleton. In the field of synthetic biology, cell membranes can be artificially reassembled.

History

Robert Hooke's discovery of cells in 1665 led to the proposal of the cell theory. Initially it was believed that all cells contained a hard cell wall since only plant cells could be observed at the time. Microscopists focused on the cell wall for well over 150 years until advances in microscopy were made. In the early 19th century, cells were recognized as being separate entities, unconnected, and bound by individual cell walls after it was found that plant cells could be separated. This theory extended to include animal cells to suggest a universal mechanism for cell protection and development.

By the second half of the 19th century, microscopy was still not advanced enough to make a distinction between cell membranes and cell walls. However, some microscopists correctly identified at this time that while invisible, it could be inferred that cell membranes existed in animal cells due to intracellular movement of components internally but not externally and that membranes were not the equivalent of a plant cell wall. It was also inferred that cell membranes were not vital components to all cells. Many refuted the existence of a cell membrane still towards the end of the 19th century. In 1890, an update to the Cell Theory stated that cell membranes existed, but were merely secondary structures. It was not until later studies with osmosis and permeability that cell membranes gained more recognition. In 1895, Ernest Overton proposed that cell membranes were made of lipids.

The lipid bilayer hypothesis, proposed in 1925 by Gorter and Grendel, created speculation in the description of the cell membrane bilayer structure based on crystallographic studies and soap bubble observations. In an attempt to accept or reject the hypothesis, researchers measured membrane thickness. These researchers extracted the lipid from human red blood cells and measured the amount of surface area the lipid would cover when spread over the surface of the water. Since mature mammalian red blood cells lack both nuclei and cytoplasmic organelles, the plasma membrane is the only lipid-containing structure in the cell. Consequently, all of the lipids extracted from the cells can be assumed to have resided in the cells' plasma membranes. The ratio of the surface area of water covered by the extracted lipid to the surface area calculated for the red blood cells from which the lipid was 2:1(approx) and they concluded that the plasma membrane contains a lipid bilayer.

In 1925 it was determined by Fricke that the thickness of erythrocyte and yeast cell membranes ranged between 3.3 and 4 nm, a thickness compatible with a lipid monolayer. The choice of the dielectric constant used in these studies was called into question but future tests could not disprove the results of the initial experiment. Independently, the leptoscope was invented in order to measure very thin membranes by comparing the intensity of light reflected from a sample to the intensity of a membrane standard of known thickness. The instrument could resolve thicknesses that depended on pH measurements and the presence of membrane proteins that ranged from 8.6 to 23.2 nm, with the lower measurements supporting the lipid bilayer hypothesis. Later in the 1930s, the membrane structure model developed in general agreement to be the paucimolecular model of Davson and Danielli (1935).

This model was based on studies of surface tension between oils and echinoderm eggs. Since the surface tension values appeared to be much lower than would be expected for an oil–water interface, it was assumed that some substance was responsible for lowering the interfacial tensions in the surface of cells. It was suggested that a lipid bilayer was in between two thin protein layers. The paucimolecular model immediately became popular and it dominated cell membrane studies for the following 30 years, until it became rivaled by the fluid mosaic model of Singer and Nicolson (1972).

Despite the numerous models of the cell membrane proposed prior to the fluid mosaic model, it remains the primary archetype for the cell membrane long after its inception in the 1970s. Although the fluid mosaic model has been modernized to detail contemporary discoveries, the basics have remained constant: the membrane is a lipid bilayer composed of hydrophilic exterior heads and a hydrophobic interior where proteins can interact with hydrophilic heads through polar interactions, but proteins that span the bilayer fully or partially have hydrophobic amino acids that interact with the non-polar lipid interior. The fluid mosaic model not only provided an accurate representation of membrane mechanics, it enhanced the study of hydrophobic forces, which would later develop into an essential descriptive limitation to describe biological macromolecules.

For many centuries, the scientists cited disagreed with the significance of the structure they were seeing as the cell membrane. For almost two centuries, the membranes were seen but mostly disregarded as an important structure with cellular function. It was not until the 20th century that the significance of the cell membrane as it was acknowledged.

Finally, two scientists Gorter and Grendel (1925) made the discovery that the membrane is "lipid-based". From this, they furthered the idea that this structure would have to be in a formation that mimicked layers. Once studied further, it was found by comparing the sum of the cell surfaces and the surfaces of the lipids, a 2:1 ratio was estimated; thus, providing the first basis of the bilayer structure known today. This discovery initiated many new studies that arose globally within various fields of scientific studies, confirming that the structure and functions of the cell membrane are widely accepted.

The structure has been variously referred to by different writers as the ectoplast (de Vries, 1885), *Plasmahaut* (plasma skin, Pfeffer, 1877, 1891), *Hautschicht* (skin layer, Pfeffer, 1886; used with a different meaning by Hofmeister, 1867), plasmatic membrane (Pfeffer, 1900), plasma membrane, cytoplasmic membrane, cell envelope and cell membrane. Some authors who did not believe that there was a functional permeable boundary at the surface of the cell preferred to use the term plasmalemma (coined by Mast, 1924) for the external region of the cell.

Composition

Cell membranes contain a variety of biological molecules, notably lipids and proteins. Composition is not set, but constantly changing for fluidity and changes in the environment, even fluctuating during different stages of cell development. Specifically, the amount of cholesterol in human primary neuron cell membrane changes, and this change in composition affects fluidity throughout development stages.

Material is incorporated into the membrane, or deleted from it, by a variety of mechanisms:

- Fusion of intracellular vesicles with the membrane (exocytosis) not only excretes the contents of the vesicle but also incorporates the vesicle membrane's components into the cell membrane. The membrane may form blebs around extracellular material that pinch off to become vesicles (endocytosis).
- If a membrane is continuous with a tubular structure made of membrane material, then material from the tube can be drawn into the membrane continuously.
- Although the concentration of membrane components in the aqueous phase is low (stable membrane components have low solubility in water), there is an exchange of molecules between the lipid and aqueous phases.

Lipids

The cell membrane consists of three classes of amphipathic lipids: phospholipids, glycolipids, and sterols. The amount of each depends upon the type of cell, but in the majority of cases phospholipids are the most abundant, often contributing for over 50% of all lipids in plasma membranes.

Glycolipids only account for a minute amount of about 2% and sterols make up the rest. In red blood cell studies, 30% of the plasma membrane is lipid. However, for the majority of eukaryotic cells, the composition of plasma membranes is about half lipids and half proteins by weight.

The fatty chains in phospholipids and glycolipids usually contain an even number of carbon atoms, typically between 16 and 20. The 16- and 18-carbon fatty acids are the most common. Fatty acids may be saturated or unsaturated, with the configuration of the double bonds nearly always "cis".

The length and the degree of unsaturation of fatty acid chains have a profound effect on membrane fluidity as unsaturated lipids create a kink, preventing the fatty acids from packing together as tightly, thus decreasing the melting temperature (increasing the fluidity) of the membrane.

The ability of some organisms to regulate the fluidity of their cell membranes by altering lipid composition is called homeoviscous adaptation.

The entire membrane is held together via non-covalent interaction of hydrophobic tails, however the structure is quite fluid and not fixed rigidly in place. Under physiological conditions phospholipid molecules in the cell membrane are in the liquid crystalline state.

It means the lipid molecules are free to diffuse and exhibit rapid lateral diffusion along the layer in which they are present. However, the exchange of phospholipid molecules between intracellular and extracellular leaflets of the bilayer is a very slow process.

Lipid rafts and caveolae are examples of cholesterol-enriched microdomains in the cell membrane. Also, a fraction of the lipid in direct contact with integral membrane proteins, which is tightly bound to the protein surface is called annular lipid shell; it behaves as a part of protein complex.

Cholesterol is normally found dispersed in varying degrees throughout cell membranes, in the irregular spaces between the hydrophobic tails of the membrane lipids, where it confers a stiffening and strengthening effect on the membrane.

Additionally, the amount of cholesterol in biological membranes varies between organisms, cell types, and even in individual cells. Cholesterol, a major component of plasma membranes, regulates the fluidity of the overall membrane, meaning that cholesterol controls the amount of movement of the various cell membrane components based on its concentrations.

In high temperatures, cholesterol inhibits the movement of phospholipid fatty acid chains, causing a reduced permeability to small molecules and reduced membrane fluidity. The opposite is true for the role of cholesterol in cooler temperatures. Cholesterol production, and thus concentration, is up-regulated (increased) in response to cold temperature.

At cold temperatures, cholesterol interferes with fatty acid chain interactions. Acting as antifreeze, cholesterol maintains the fluidity of the membrane. Cholesterol is more abundant in cold-weather animals than warm-weather animals. In plants, which lack cholesterol, related compounds called sterols perform the same function as cholesterol.

Phospholipids Forming Lipid Vesicles

Lipid vesicles or liposomes are approximately spherical pockets that are enclosed by a lipid bilayer. These structures are used in laboratories to study the effects of chemicals in cells by delivering these chemicals directly to the cell, as well as getting more insight into cell membrane permeability. Lipid vesicles and liposomes are formed by first suspending a lipid in an aqueous solution then agitating the mixture through sonication, resulting in a vesicle.

Measuring the rate of efflux from the inside of the vesicle to the ambient solution allows researchers to better understand membrane permeability. Vesicles can be formed with molecules and ions inside the vesicle by forming the vesicle with the desired molecule or ion present in the solution. Proteins can also be embedded into the membrane through solubilizing the desired proteins in the presence of detergents and attaching them to the phospholipids in which the liposome is formed. These provide researchers with a tool to examine various membrane protein functions.

Carbohydrates

Plasma membranes also contain carbohydrates, predominantly glycoproteins, but with some glycolipids (cerebrosides and gangliosides). Carbohydrates are important in the role of cell-cell recognition in eukaryotes; they are located on the surface of the cell where they recognize host cells and share information. Viruses that bind to cells using these receptors cause an infection. For the most part, no glycosylation occurs on membranes within the cell; rather generally glycosylation occurs on the extracellular surface of the plasma membrane. The glycocalyx is an important feature in all cells, especially epithelia with microvilli. Recent data suggest the glycocalyx participates in cell adhesion, lymphocyte homing, and many others. The penultimate sugar is galactose and the terminal sugar is sialic acid, as the sugar backbone is modified in the Golgi apparatus. Sialic acid carries a negative charge, providing an external barrier to charged particles.

Proteins

The cell membrane has large content of proteins, typically around 50% of membrane volume These proteins are important for the cell because they are responsible for various biological activities. Approximately a third of the genes in yeast code specifically for them, and this number is even higher in multicellular organisms.

Membrane proteins consist of three main types: integral proteins, peripheral proteins, and lipid-anchored proteins.

Type	Description	Examples
Integral proteins or transmembrane proteins	Span the membrane and have a hydrophilic cytosolic domain, which interacts with internal molecules, a hydrophobic membrane-spanning domain that anchors it within the cell membrane, and a hydrophilic extracellular domain that interacts with external molecules. The hydrophobic domain consists of one, multiple, or a combination of α-helices and β sheet protein motifs.	Ion channels, proton pumps, G protein-coupled receptor
Lipid anchored proteins	Covalently bound to single or multiple lipid molecules; hydrophobically insert into the cell membrane and anchor the protein. The protein itself is not in contact with the membrane.	G proteins
Peripheral proteins	Attached to integral membrane proteins, or associated with peripheral regions of the lipid bilayer. These proteins tend to have only temporary interactions with biological membranes, and once reacted, the molecule dissociates to carry on its work in the cytoplasm.	Some enzymes, some hormones

Integral proteins are amphipathic transmembrane proteins. Examples of integral proteins include ion channels, proton pumps, and g-protein coupled receptors. Ion channels allow inorganic ions such as sodium, potassium, calcium, or chlorine to diffuse down their electrochemical gradient across the lipid bilayer through hydrophilic pores across the membrane. The electrical behavior of cells (i.e. nerve cells) is controlled by ion channels. Proton pumps are protein pumps that are embedded in the lipid bilayer that allow protons to travel through the membrane by transferring from one amino acid side chain to another. Processes such as electron transport and generating ATP use proton pumps. A G-protein coupled receptor is a single polypeptide chain that crosses the lipid bilayer seven times responding to signal molecules (i.e. hormones and neurotransmitters). G-protein coupled receptors are used in processes such as cell to cell signaling, the regulation of the production of cAMP, and the regulation of ion channels.

The cell membrane, being exposed to the outside environment, is an important site of cell–cell communication. As such, a large variety of protein receptors and identification proteins, such as antigens, are present on the surface of the membrane. Functions of membrane proteins can also include cell–cell contact, surface recognition, cytoskeleton contact, signaling, enzymatic activity, or transporting substances across the membrane.

Most membrane proteins must be inserted in some way into the membrane. For this to occur, an N-terminus "signal sequence" of amino acids directs proteins to the endoplasmic reticulum, which inserts the proteins into a lipid bilayer. Once inserted, the proteins are then transported to their final destination in vesicles, where the vesicle fuses with the target membrane.

Function

The cell membrane surrounds the cytoplasm of living cells, physically separating the intracellular components from the extracellular environment. The cell membrane also plays a role in anchoring the cytoskeleton to provide shape to the cell, and in attaching to the extracellular matrix and other cells to hold them together to form tissues. Fungi, bacteria, most archaea, and plants also have a cell wall, which provides a mechanical support to the cell and precludes the passage of larger molecules.

The cell membrane is selectively permeable and able to regulate what enters and exits the cell, thus facilitating the transport of materials needed for survival. The movement of substances across the membrane can be achieved by either passive transport, occurring without the input of cellular energy, or by active transport, requiring the cell to expend energy in transporting it. The membrane also maintains the cell potential. The cell membrane thus works as a selective filter that allows only certain things to come inside or go outside the cell. The cell employs a number of transport mechanisms that involve biological membranes:

Passive Osmosis and Diffusion

Some substances (small molecules, ions) such as carbon dioxide (CO_2) and oxygen (O_2), can move across the plasma membrane by diffusion, which is a passive transport process. Because the membrane acts as a barrier for certain molecules and ions, they can occur in different concentrations on the two sides of the membrane. Diffusion occurs when small molecules and ions move freely from high concentration to low concentration in order to equilibrate the membrane. It is considered a passive transport process because it does not require energy and is propelled by the concentration gradient created by each side of the membrane. Such a concentration gradient across a semipermeable membrane sets up an osmotic flow for the water. Osmosis, in biological systems involves a solvent, moving through a semipermeable membrane similarly to passive diffusion as the solvent still moves with the concentration gradient and requires no energy. While water is the most common solvent in cell, it can also be other liquids as well as supercritical liquids and gases.

Transmembrane Protein Channels and Transporters

Transmembrane proteins extend through the lipid bilayer of the membranes; they function on both sides of the membrane to transport molecules across it. Nutrients, such as sugars or amino acids, must enter the cell, and certain products of metabolism must leave the cell. Such molecules can diffuse passively through protein channels such as aquaporins in facilitated diffusion or are pumped across the membrane by transmembrane transporters. Protein channel proteins, also called *permeases*, are usually quite specific, and they only recognize and transport a limited variety of chemical substances, often limited to a single substance. Another example of a transmembrane protein is a cell-surface receptor, which allow cell signaling molecules to communicate between cells.

Endocytosis

Endocytosis is the process in which cells absorb molecules by engulfing them. The plasma membrane creates a small deformation inward, called an invagination, in which the substance to be transported is captured. This invagination is caused by proteins on the outside on the cell membrane, acting as receptors and clustering into depressions that eventually promote accumulation of more proteins and lipids on the cytosolic side of the membrane. The deformation then pinches off from the membrane on the inside of the cell, creating a vesicle containing the captured substance. Endocytosis is a pathway for internalizing solid particles ("cell eating" or phagocytosis), small molecules and ions ("cell drinking" or pinocytosis), and macromolecules. Endocytosis requires energy and is thus a form of active transport.

Exocytosis

Just as material can be brought into the cell by invagination and formation of a vesicle, the membrane of a vesicle can be fused with the plasma membrane, extruding its contents to the surrounding medium. This is the process of exocytosis. Exocytosis occurs in various cells to remove undigested residues of substances brought in by endocytosis, to secrete substances such as hormones and enzymes, and to transport a substance completely across a cellular barrier. In the process of exocytosis, the undigested waste-containing food vacuole or the secretory vesicle budded from Golgi apparatus, is first moved by cytoskeleton from the interior of the cell to the surface. The vesicle membrane comes in contact with the plasma membrane. The lipid molecules of the two bilayers rearrange themselves and the two membranes are, thus, fused. A passage is formed in the fused membrane and the vesicles discharges its contents outside the cell.

Prokaryotes

Prokaryotes are divided into two different groups, Archaea and Bacteria, with bacteria dividing further into gram-positive and gram-negative. Gram-negative bacteria have both a plasma membrane and an outer membrane separated by periplasm; however, other prokaryotes have only a plasma membrane. These two membranes differ in many aspects. The outer membrane of the gram-negative bacteria differs from other prokaryotes due to phospholipids forming the exterior of the bilayer, and lipoproteins and phospholipids forming the interior. The outer membrane typically has a porous quality due to its presence of membrane proteins, such as gram-negative porins, which are pore-forming proteins. The inner plasma membrane is also generally symmetric whereas the outer membrane is asymmetric because of proteins such as the aforementioned.

Also, for the prokaryotic membranes, there are multiple things that can affect the fluidity. One of the major factors that can affect the fluidity is fatty acid composition. For example, when the bacteria *Staphylococcus aureus* was grown in 37°C for 24h, the membrane exhibited a more fluid state instead of a gel-like state. This supports the concept that in higher temperatures, the membrane is more fluid than in colder temperatures. When the membrane is becoming more fluid and needs to become more stabilized, it will make longer fatty acid chains or saturated fatty acid chains in order to help stabilize the membrane.

Bacteria are also surrounded by a cell wall composed of peptidoglycan (amino acids and sugars). Some eukaryotic cells also have cell walls, but none that are made of peptidoglycan. The outer membrane of gram negative bacteria is rich in lipopolysaccharides, which are combined poly- or oligosaccharide and carbohydrate lipid regions that stimulate the cell's natural immunity. The outer membrane can bleb out into periplasmic protrusions under stress conditions or upon virulence requirements while encountering a host target cell, and thus such blebs may work as virulence organelles. Bacterial cells provide numerous examples of the diverse ways in which prokaryotic cell membranes are adapted with structures that suit the organism's niche. For example, proteins on the surface of certain bacterial cells aid in their gliding motion. Many gram-negative bacteria have cell membranes which contain ATP-driven protein exporting systems.

Structures

Fluid Mosaic Model

According to the fluid mosaic model of S. J. Singer and G. L. Nicolson (1972), which replaced the earlier model of Davson and Danielli, biological membranes can be considered as a two-dimensional liquid in which lipid and protein molecules diffuse more or less easily. Although the lipid bilayers that form the basis of the membranes do indeed form two-dimensional liquids by themselves, the plasma membrane also contains a large quantity of proteins, which provide more structure. Examples of such structures are protein-protein complexes, pickets and fences formed by the actin-based cytoskeleton, and potentially lipid rafts.

Lipid Bilayer

Lipid bilayers form through the process of self-assembly. The cell membrane consists primarily of a thin layer of amphipathic phospholipids that spontaneously arrange so that the hydrophobic "tail" regions are isolated from the surrounding water while the hydrophilic "head" regions interact with the intracellular (cytosolic) and extracellular faces of the resulting bilayer. This forms a continuous, spherical lipid bilayer. Hydrophobic interactions (also known as the hydrophobic effect) are the major driving forces in the formation of lipid bilayers. An increase in interactions between hydrophobic molecules (causing clustering of hydrophobic regions) allows water molecules to bond more freely with each other, increasing the entropy of the system. This complex interaction can include noncovalent interactions such as van der Waals, electrostatic and hydrogen bonds.

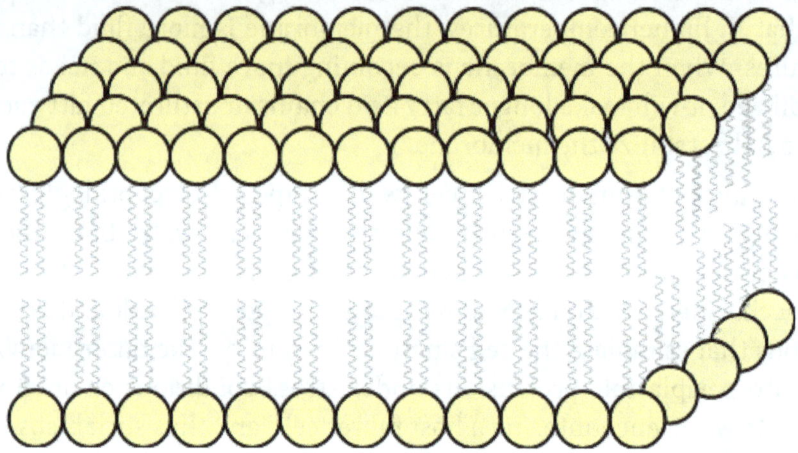

Fig 3: Diagram of the Arrangement of Amphipathic Lipid Molecules to Form a Lipid Bilayer. The Yellow Polar Head Groups Separate the Grey Hydrophobic Tails from the Aqueous Cytosolic and Extracellular Environments

Lipid bilayers are generally impermeable to ions and polar molecules. The arrangement of hydrophilic heads and hydrophobic tails of the lipid bilayer prevent polar solutes (ex. amino acids, nucleic acids, carbohydrates, proteins, and ions) from diffusing across the membrane, but generally allows for the passive diffusion of hydrophobic molecules. This affords the cell the ability to control the movement of these substances via transmembrane protein complexes such as pores, channels and gates. Flippases and scramblases concentrate phosphatidyl serine, which carries a negative charge, on the inner membrane. Along with NANA, this creates an extra barrier to charged moieties moving through the membrane.

Membranes serve diverse functions in eukaryotic and prokaryotic cells. One important role is to regulate the movement of materials into and out of cells. The phospholipid bilayer structure (fluid mosaic model) with specific membrane proteins accounts for the selective permeability of the membrane and passive and active transport mechanisms. In addition, membranes in prokaryotes and in the mitochondria and chloroplasts of eukaryotes facilitate the synthesis of ATP through chemiosmosis.

Membrane Polarity

The apical membrane or luminal membrane of a polarized cell is the surface of the plasma membrane that faces inward to the lumen. This is particularly evident in epithelial and endothelial cells, but also describes other polarized cells, such as neurons. The basolateral membrane or basolateral cell membrane of a polarized cell is the surface of the plasma membrane that forms its basal and lateral surfaces. It faces outwards, towards the interstitium, and away from the lumen. Basolateral membrane is a compound phrase referring to the terms "basal (base) membrane" and "lateral (side) membrane", which, especially in epithelial cells, are identical in composition and activity. Proteins (such as ion channels and pumps) are free to move from the basal to the lateral surface of the cell or vice versa in accordance with the fluid mosaic model. Tight junctions join epithelial cells near their apical surface to prevent the migration of proteins from the basolateral membrane to the apical membrane. The basal and lateral surfaces thus remain roughly equivalent to one another, yet distinct from the apical surface.

Membrane Structures

Cell membrane can form different types of "supramembrane" structures such as caveolae, postsynaptic density, podosomes, invadopodia, focal adhesion, and different types of cell junctions. These structures are usually responsible for cell adhesion, communication, endocytosis and exocytosis. They can be visualized by electron microscopy or fluorescence microscopy. They are composed of specific proteins, such as integrins and cadherins.

Cytoskeleton

The cytoskeleton is found underlying the cell membrane in the cytoplasm and provides a scaffolding for membrane proteins to anchor to, as well as forming organelles that extend from the cell. Indeed, cytoskeletal elements interact extensively and intimately with the cell membrane. Anchoring proteins restricts them to a particular cell surface — for example, the apical surface of epithelial cells that line the vertebrate gut — and limits how far they may diffuse within the bilayer.

The cytoskeleton is able to form appendage-like organelles, such as cilia, which are microtubule-based extensions covered by the cell membrane, and filopodia, which are actin-based extensions. These extensions are ensheathed in membrane and project from the surface of the cell in order to sense the external environment and/or make contact with the substrate or other cells. The apical surfaces of epithelial cells are dense with actin-based finger-like projections known as microvilli, which increase cell surface area and thereby increase the absorption rate of nutrients. Localized decoupling of the cytoskeleton and cell membrane results in formation of a bleb.

Intracellular Membranes

The content of the cell, inside the cell membrane, is composed of numerous membrane-bound organelles, which contribute to the overall function of the cell. The origin, structure, and function of each organelle leads to a large variation in the cell composition due to the individual uniqueness associated with each organelle.

- Mitochondria and chloroplasts are considered to have evolved from bacteria, known as the endosymbiotic theory. This theory arose from the idea that *Paracoccus* and *Rhodopseudomonas*, types of bacteria, share similar functions to mitochondria and blue-green algae (cyanobacteria) share similar functions to chloroplasts. Endosymbiotic theory proposes that through the course of evolution, a eukaryotic cell engulfed these two types of bacteria, leading to the formation of mitochondria and chloroplasts inside eukaryotic cells. This engulfment lead to the double-membranes systems of these organelles in which the outer membrane originated from the host's plasma membrane and the inner membrane was the endosymbiont's plasma membrane. Considering that mitochondria and chloroplasts both contain their own DNA is further support that both of these organelles evolved from engulfed bacteria that thrived inside a eukaryotic cell.
- In eukaryotic cells, the nuclear membrane separates the contents of the nucleus from the cytoplasm of the cell. The nuclear membrane is formed by an inner and outer membrane, providing the strict regulation of materials in to and out of the nucleus. Materials move between the cytosol and the nucleus through nuclear pores in the nuclear membrane. If a cell's nucleus is more active in transcription, its membrane will have more pores. The protein composition of the nucleus can vary greatly from the cytosol as many proteins are unable to cross through pores via diffusion. Within the nuclear membrane, the inner and outer membranes vary in protein composition, and only the outer membrane is continuous with the endoplasmic reticulum (ER) membrane. Like the ER, the outer membrane also possesses ribosomes responsible for producing and transporting proteins into the space between the two membranes. The nuclear membrane disassembles during the early stages of mitosis and reassembles in later stages of mitosis.
- The ER, which is part of the endomembrane system, which makes up a very large portion of the cell's total membrane content. The ER is an enclosed network of tubules and sacs, and its main functions include protein synthesis, and lipid metabolism. There are 2 types of ER, smooth and rough. The rough ER has ribosomes attached to it used for protein

synthesis, while the smooth ER is used more for the processing of toxins and calcium regulation in the cell.

- The Golgi apparatus has two interconnected round Golgi cisternae. Compartments of the apparatus forms multiple tubular-reticular networks responsible for organization, stack connection and cargo transport that display a continuous grape-like stringed vesicles ranging from 50 to 60 nm. The apparatus consists of three main compartments, a flat disc-shaped cisterna with tubular-reticular networks and vesicles.

Variations

The cell membrane has different lipid and protein compositions in distinct types of cells and may have therefore specific names for certain cell types.

- **Sarcolemma in Muscle Cells:** Sarcolemma is the name given to the cell membrane of muscle cells. Although the sarcolemma is similar to other cell membranes, it has other functions that set it apart. For instance, the sarcolemma transmits synaptic signals, helps generate action potentials, and is very involved in muscle contraction. Unlike other cell membranes, the sarcolemma makes up small channels called T-tubules that pass through the entirety of muscle cells. It has also been found that the average sarcolemma is 10 nm thick as opposed to the 4 nm thickness of a general cell membrane.
- **Oolemma is the Cell Membrane in Oocytes:** The oolemma of oocytes, (immature egg cells) are not consistent with a lipid bilayer as they lack a bilayer and do not consist of lipids. Rather, the structure has an inner layer, the fertilization envelope, and the exterior is made up of the vitelline layer, which is made up of glycoproteins; however, channels and proteins are still present for their functions in the membrane.
- **Axolemma:** The specialized plasma membrane on the axons of nerve cells that is responsible for the generation of the action potential. It consists of a granular, densely packed lipid bilayer that works closely with the cytoskeleton components spectrin and actin. These cytoskeleton components are able to bind to and interact with transmembrane proteins in the axolemma.

Permeability

The permeability of a membrane is the rate of passive diffusion of molecules through the membrane. These molecules are known as permeant molecules. Permeability depends mainly on the electric charge and polarity of the molecule and to a lesser extent the molar mass of the molecule. Due to the cell membrane's hydrophobic nature, small electrically neutral molecules pass through the membrane more easily than charged, large ones. The inability of charged molecules to pass through the cell membrane results in pH partition of substances throughout the fluid compartments of the body.

3.1.3 Cytoplasm

In cell biology, the cytoplasm describes all material within a eukaryotic cell, enclosed by the cell membrane, except for the cell nucleus. The material inside the nucleus and contained within the nuclear membrane is termed the nucleoplasm. The main components of the cytoplasm are the cytosol (a gel-like substance), the organelles (the cell's internal sub-structures), and various cytoplasmic inclusions. The cytoplasm is about 80% water and is usually colorless.

The submicroscopic ground cell substance, or cytoplasmic matrix, that remains after the exclusion of the cell organelles and particles is groundplasm. It is the hyaloplasm of light microscopy, a highly complex, polyphasic system in which all resolvable cytoplasmic elements are suspended, including the larger organelles such as the ribosomes, mitochondria, plant plastids, lipid droplets, and vacuoles.

Many cellular activities take place within the cytoplasm, such as many metabolic pathways, including glycolysis, photosynthesis, and processes such as cell division. The concentrated inner area is called the endoplasm and the outer layer is called the cell cortex, or ectoplasm. Movement of calcium ions in and out of the cytoplasm is a signaling activity for metabolic processes. In plants, movement of the cytoplasm around vacuoles is known as cytoplasmic streaming.

History

The term was introduced by Rudolf von Kölliker in 1863, originally as a synonym for protoplasm, but later it has come to mean the cell substance and organelles outside the nucleus. There has been certain disagreement on the definition of cytoplasm, as some authors prefer to exclude from it some organelles, especially the vacuoles and sometimes the plastids.

Physical Nature

It remains uncertain how the various components of the cytoplasm interact to allow movement of organelles while maintaining the cell's structure. The flow of cytoplasmic components plays an important role in many cellular functions which are dependent on the permeability of the cytoplasm.

An example of such function is cell signalling, a process which is dependent on the manner in which signaling molecules are allowed to diffuse across the cell. While small signaling molecules like calcium ions are able to diffuse with ease, larger molecules and subcellular structures often require aid in moving through the cytoplasm.

The irregular dynamics of such particles have given rise to various theories on the nature of the cytoplasm.

As a Sol-Gel

There has long been evidence that the cytoplasm behaves like a sol-gel. It is thought that the component molecules and structures of the cytoplasm behave at times like a disordered colloidal solution (sol) and at other times like an integrated network, forming a solid mass (gel). This theory thus proposes that the cytoplasm exists in distinct fluid and solid phases depending on the level of interaction between cytoplasmic components, which may explain the differential dynamics of different particles observed moving through the cytoplasm. A papers suggested that at length scale smaller than 100 nm, the cytoplasm acts like a liquid, while in a larger length scale, it acts like a gel.

As a Glass

It has been proposed that the cytoplasm behaves like a glass-forming liquid approaching the glass transition. In this theory, the greater the concentration of cytoplasmic components, the less the cytoplasm behaves like a liquid and the more it behaves as a solid glass, freezing more significant cytoplasmic components in place (it is thought that the cell's metabolic activity can fluidize the cytoplasm to allow the movement of such more significant cytoplasmic components). A cell's ability to vitrify in the absence of metabolic activity, as in dormant periods, may be beneficial as a defense strategy. A solid glass cytoplasm would freeze subcellular structures in place, preventing damage, while allowing the transmission of tiny proteins and metabolites, helping to kickstart growth upon the cell's revival from dormancy.

Other Perspectives

Research has examined the motion of cytoplasmic particles independent of the nature of the cytoplasm. In such an alternative approach, the aggregate random forces within the cell caused by motor proteins explain the non-Brownian motion of cytoplasmic constituents.

Constituents

Cytosol

The cytosol is the portion of the cytoplasm not contained within membrane-bound organelles. Cytosol makes up about 70% of the cell volume and is a complex mixture of cytoskeleton filaments, dissolved molecules, and water. The cytosol's filaments include the protein filaments such as actin filaments and microtubules that make up the cytoskeleton, as well as soluble proteins and small structures such as ribosomes, proteasomes, and the mysterious vault complexes. The inner, granular and more fluid portion of the cytoplasm is referred to as endoplasm. Due to this network of fibres and high concentrations of dissolved macromolecules, such as proteins, an effect called macromolecular crowding occurs and the cytosol does not act as an ideal solution. This crowding effect alters how the components of the cytosol interact with each other.

Organelles

Organelles (literally "little organs") are usually membrane-bound structures inside the cell that have specific functions. Some major organelles that are suspended in the cytosol are the mitochondria, the endoplasmic reticulum, the Golgi apparatus, vacuoles, lysosomes, and in plant cells, chloroplasts.

Cytoplasmic Inclusions

The inclusions are small particles of insoluble substances suspended in the cytosol. A huge range of inclusions exist in different cell types, and range from crystals of calcium oxalate or silicon dioxide in plants, to granules of energy-storage materials such as starch, glycogen, or polyhydroxybutyrate. A particularly widespread example are lipid droplets, which are spherical droplets composed of lipids and proteins that are used in both prokaryotes and eukaryotes as a way of storing lipids such as fatty acids and sterols. Lipid droplets make up much of the volume of adipocytes, which are specialized lipid-storage cells, but they are also found in a range of other cell types.

Controversy and Research

The cytoplasm, mitochondria, and most organelles are contributions to the cell from the maternal gamete. Contrary to the older information that disregards any notion of the cytoplasm being active, new research has shown it to be in control of movement and flow of nutrients in and out of the cell by viscoplastic behavior and a measure of the reciprocal rate of bond breakage within the cytoplasmic network.

The material properties of the cytoplasm remain an ongoing investigation. A method of determining the mechanical behaviour of living cell mammalian cytoplasm with the aid of optical tweezers has been described.

3.2 CLASSIFICATION OF BACTERIA

Bacteria are classified based on several criteria, including their shape, cell wall structure, and metabolic needs.

- **Shape:** Bacteria come in various shapes, including spherical (cocci), rod-shaped (bacilli), spiral (spirilla), and corkscrew-shaped (spirochetes). These shapes help in identifying bacterial species and understanding their modes of life.

- **Gram Staining:** The Gram stain differentiates bacteria into Gram-positive and Gram-negative groups. Gram-positive bacteria have a thick cell wall that retains the violet color of the stain, while Gram-negative bacteria have a thinner cell wall with an outer membrane and do not retain the violet stain.
- **Oxygen Requirement:** Bacteria can be classified as aerobic (requiring oxygen), anaerobic (not requiring oxygen), or facultative anaerobes (able to grow with or without oxygen). This classification is important in understanding bacterial environments and their potential applications or hazards.
- **Metabolic Pathways:** Bacteria use various metabolic pathways, including photosynthesis, nitrogen fixation, and fermentation. Some bacteria, like cyanobacteria, are photosynthetic and contribute significantly to oxygen production on Earth, while others thrive in low-oxygen environments and use alternative energy sources.

3.3 THE ROLE OF BACTERIA IN THE ENVIRONMENT

Bacteria play a crucial role in sustaining life and maintaining ecological balance. Their functions are fundamental in natural processes like nutrient cycling, waste decomposition, and even climate regulation.

- **Nutrient Cycling:** Bacteria are essential for nutrient cycling, including the nitrogen, sulfur, and carbon cycles. For example, nitrogen-fixing bacteria convert atmospheric nitrogen into forms that plants can absorb and use, promoting soil fertility.
- **Decomposition:** Bacteria break down organic matter, such as dead plants and animals, turning it into simpler substances that plants can reuse. This decomposition is critical for soil health and plant growth.
- **Bioremediation:** Certain bacteria can degrade pollutants and toxins in the environment. They are used in bioremediation to clean up oil spills, treat sewage, and reduce pollution. Bacteria's ability to break down complex compounds makes them valuable for environmental restoration.
- **Methane Regulation:** Methanotrophic bacteria consume methane, a potent greenhouse gas, and help reduce its concentration in the atmosphere. These bacteria play a role in mitigating climate change impacts by preventing excessive methane release.

3.4 BACTERIA AND HUMAN HEALTH

Bacteria are often associated with infectious diseases, but they also have many beneficial roles in human health. The human body hosts trillions of bacterial cells, many of which are essential for proper functioning.

- **Gut Microbiome**: The human gut contains a complex community of bacteria, known as the gut microbiome, that aids digestion, synthesizes vitamins, and supports immune health. Beneficial bacteria in the gut produce short-chain fatty acids that promote gut health and prevent the growth of harmful bacteria.
- **Probiotics and Health Benefits:** Some bacteria, known as probiotics, are deliberately ingested to support a healthy balance of gut bacteria. Probiotics are thought to improve digestive health, boost immunity, and even influence mental well-being.
- **Pathogenic Bacteria**: Not all bacteria are beneficial. Pathogenic bacteria cause diseases like tuberculosis, cholera, and typhoid. They can be spread through contaminated food, water, or direct contact. Medical science has developed antibiotics to combat bacterial infections, though antibiotic resistance has become a serious issue.
- **Antibiotic Resistance**: Bacteria can evolve rapidly, leading to antibiotic resistance. Overuse and misuse of antibiotics have accelerated this process, making certain bacterial infections harder to treat. Antibiotic-resistant strains, such as MRSA (Methicillin-resistant *Staphylococcus aureus*), pose significant challenges to healthcare systems worldwide.

3.5 INDUSTRIAL AND BIOTECHNOLOGICAL APPLICATIONS OF BACTERIA

Bacteria are invaluable in industries ranging from food production to pharmaceuticals and environmental management. Their ability to produce specific enzymes, proteins, and other compounds has opened up numerous applications.

- **Food Production**: Bacteria are used in fermentation to produce foods like yogurt, cheese, and pickles. Fermentative bacteria break down sugars into lactic acid, preserving food and adding unique flavors.
- **Pharmaceuticals**: Bacteria are harnessed in pharmaceutical production, such as the creation of antibiotics, enzymes, and vaccines. Genetic engineering techniques enable scientists to produce human insulin and other critical drugs using bacterial cells as factories.
- **Bioengineering and Genetic Manipulation**: Bacteria have been genetically modified for various purposes, including the production of biofuels, biodegradable plastics, and other renewable resources. The

simplicity of bacterial cells makes them ideal for genetic experiments and applications in synthetic biology.

- **Agriculture:** Beneficial bacteria, including nitrogen-fixing bacteria and phosphate-solubilizing bacteria, improve soil fertility and support plant growth. Some bacteria are used as biopesticides to control agricultural pests, reducing the need for chemical pesticides.

3.6 CHALLENGES AND FUTURE DIRECTIONS IN BACTERIAL RESEARCH

The study of bacteria has come a long way, but it also presents significant challenges and opportunities for future research.

- **Antibiotic Resistance and Public Health:** The emergence of antibiotic-resistant bacteria remains a global health crisis. Researchers are exploring new antibiotics, bacteriophages (viruses that infect bacteria), and alternative therapies to combat resistant strains.
- **Synthetic Biology and Genetic Engineering:** Bacteria are key organisms in the field of synthetic biology, where they are engineered to perform specific tasks, such as producing biofuels or sensing environmental pollutants. The potential to reprogram bacteria could transform various industries, although ethical and safety concerns remain.
- **Microbiome Research:** The human microbiome continues to be a major research focus. Understanding the complex interactions between bacterial communities in the human body and their impact on health, behavior, and disease is a growing field with implications for personalized medicine and wellness.
- **Climate Change and Bacteria:** Bacteria will play a role in addressing environmental challenges related to climate change. Methane-consuming bacteria, nitrogen-fixing bacteria, and others could help manage greenhouse gases and improve agricultural productivity in a changing climate.

3.7 CONCLUSION

Bacteria, despite their small size and simplicity, are incredibly versatile and influential organisms that have shaped life on Earth. From their vital roles in ecosystems and human health to their industrial applications, bacteria are essential in numerous fields.

However, challenges like antibiotic resistance and the need for sustainable environmental management highlight the importance of continued research and innovation in bacterial science. Bacteria are not merely pathogens; they are also powerful allies in medicine, industry, and environmental protection, underscoring the need to understand and harness their potential responsibly.

CHAPTER 4

FUNGI

Fungi constitute one of the major kingdoms of life, displaying remarkable diversity and occupying diverse ecological roles. Though often associated with mold or mushrooms, fungi encompass a vast range of organisms that contribute significantly to ecosystems, medicine, and industry. Unlike plants, fungi lack chlorophyll and obtain their nutrients through absorption, often as decomposers.

4.1 DEFINING CHARACTERISTICS OF FUNGI

Fungi are distinct from other kingdoms of life due to several unique characteristics. They are eukaryotic organisms, meaning they have complex cell structures with membrane-bound organelles, such as a nucleus and mitochondria. However, they differ from plants, animals, and bacteria in crucial ways.

- **Cell Structure**: Fungal cells have a rigid cell wall composed mainly of chitin, a carbohydrate also found in the exoskeleton of arthropods. This cell wall provides structural support and protection.
- **Mode of Nutrition**: Fungi are heterotrophic organisms, meaning they cannot produce their own food. Instead, they absorb nutrients from their surroundings, often by secreting enzymes that break down complex molecules. Fungi can be saprophytic (feeding on dead matter), parasitic (feeding on living hosts), or mutualistic (forming symbiotic relationships).
- **Growth Form:** Fungi exist in various forms, including unicellular yeasts, filamentous molds, and macroscopic mushrooms. Many fungi grow as hyphae—long, thread-like structures that form a dense network called mycelium. This extensive network enables fungi to absorb nutrients from a large surface area.

4.2 CLASSIFICATION OF FUNGI

Fungi are classified into different groups based on their reproductive structures and life cycles. The main groups include:

4.2.1 Zygomycota

Zygomycota, or zygote fungi, is a former division or phylum of the kingdom Fungi. The members are now part of two phyla: the Mucoromycota and Zoopagomycota.

Approximately 1060 species are known. They are mostly terrestrial in habitat, living in soil or on decaying plant or animal material. Some are parasites of plants, insects, and small animals, while others form symbiotic relationships with plants. Zygomycete hyphae may be coenocytic, forming septa only where gametes are formed or to wall off dead hyphae. Zygomycota is no longer recognised as it was not believed to be truly monophyletic.

Spores

The term "spore" is used to describe a structure related to propagation and dispersal. Zygomycete spores can be formed through both sexual and asexual means. Before germination the spore is in a dormant state. During this period, the metabolic rate is very low and it may last from a few hours to many years. There are two types of dormancy.

The exogenous dormancy is controlled by environmental factors such as temperature or nutrient availability. The endogenous or constitutive dormancy depends on characteristics of the spore itself; for example, metabolic features. In this type of dormancy, germination may be prevented even if the environmental conditions favor growth.

Chlamydospores

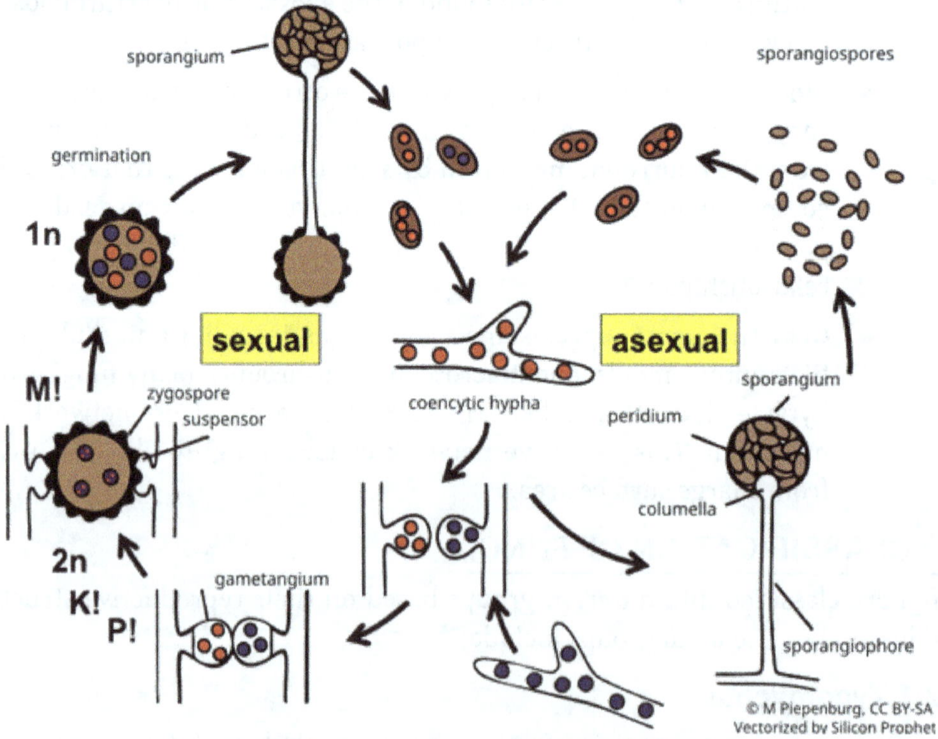

Fig 1: Zygospore Diagram and Reproduction

Chlamydospores are asexual spores different from sporangiospores. The primary function of chlamydospores is the persistence of the mycelium and they are released when the mycelium degrades. Chlamydospores have no mechanism for dispersal. In zygomycetes the formation of chlamydospores is usually intercalar. However, it may also be terminal. In accordance with their function chlamydospores have a thick cell wall and are pigmented.

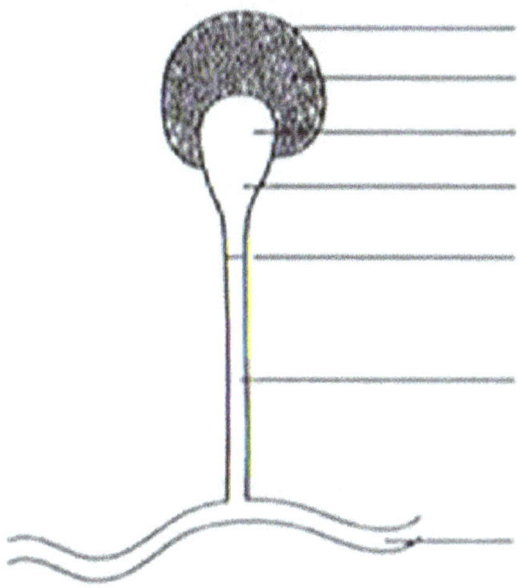

Fig 2: Sporangium

Mitospores

In zygomycetes, mitospores (sporangiospores) are formed asexually. They are formed in specialized structures, the mitosporangia (sporangia) that contain few to several thousand of spores, depending on the species. Mitosporangia are carried by specialized hyphae, the mitosporangiophores (sporangiophores). These specialized hyphae usually show negative gravitropism and positive phototropism allowing good spore dispersal.

The sporangia wall is thin and is easily destroyed by mechanical stimuli (e.g., falling raindrops, passing animals), leading to the dispersal of the ripe mitospores. The walls of these spores contain sporopollenin in some species. Sporopollenin is formed out of β-carotene and is very resistant to biological and chemical degradation. Zygomycete spores may also be classified in respect to their persistence:

Zygophores

Zygophores are chemotropic aerial hyphae that are the sex organs of zygomycota, except for Phycomyces in which they are not aerial but found in the substratum. They have two different mating types (+) and (-). The opposite mating types grow towards each other due to volatile pheromones given off by the opposite strand, mainly trisporic acid and its precursors. Once two opposite mating types have made initial contact, they give rise to a zygospore through multiple steps. Once contact between the zygophores has been made, their walls adhere to each other, flatten and then the contact site is referred to as the fusion septum. The tips of the zygophore become distended and form what is called the progametangia. A septum develops by gradual inward extension until it separates the terminal gametangia from the progametangial base.

At this point the zygophore is then called the suspensor. Vesicles accumulate at the fusion septum at which time it begins to dissolve. A little before the fusion septum completely dissolves, the primary outer wall begins to thicken. This can be seen as dark patches on the primary wall as the fusion septum dissolves. These dark patches on the wall will eventually develop into warty structures that make up the thickness of the zygospore wall. As the zygospore enlarges, so do the warty structures until there are contiguous around the entire cell. At this point, electron microscopy can no longer penetrate the wall. Eventually the warts push through the primary wall and darken which is likely caused by melanin.

Meiosis usually occurs before zygospore germination and there are a few main types of distinguishable nuclear behavior. Type 1 is when the nuclei fuse quickly, within a few days, resulting in mature zygospore having haploid nuclei. Type 2 is when some nuclei do not pair and degenerate instead, meiosis is delayed until germination.

Type 3 is when haploid nuclei continue to divide mitotically and then some associate into groups and some do not. This results in diploid and haploid nuclei being found in the germ sporangium.

Cell Wall

Fig 3: Typical Fungal Cell Wall Structure

Zygomycetes exhibit a special structure of cell wall. Most fungi have chitin as structural polysaccharide, while zygomycetes synthesize chitosan, the deacetylated homopolymer of chitin. Chitin is built of β-1,4 bonded *N*-acetyl glucosamine. Fungal hyphae grow at the tip. Therefore, specialized vesicles, the chitosomes, bring precursors of chitin and its synthesizing enzyme, chitin synthetase, to the outside of the membrane by exocytosis. The enzyme on the membrane catalyzes glycosidic bond formations from the nucleotide sugar substrate, uridine diphospho-*N*-acetyl-D-glucosamine.

The nascent polysaccharide chain is then cleaved by the enzyme chitin deacetylase. The enzyme catalyzes the hydrolytic cleavage of the *N*-acetamido group in chitin. After this the chitosan polymer chain forms micro fibrils. These fibers are embedded in an amorphous matrix consisting of proteins, glucans (which putatively cross-link the chitosan fibers), mannoproteins, lipids and other compounds.

Trisporic Acid

Trisporic acid is a C-18 terpenoid compound that is synthesized via β-carotene and retinol pathways in the zygomycetes. It is a pheromone compound responsible for sexual differentiation in those fungal species.

Trisporic acid was discovered in 1964 as a metabolite that caused enhanced carotene production in *Blakeslea trispora*. It was later shown to be the hormone that brought about zygophore production in *Mucor mucedo*. The American mycologist and geneticist Albert Francis Blakeslee discovered that some species of Mucorales were self-sterile (heterothallic), in which interactions of two strains, designated (+) and (-), are necessary for the initiation of sexual activity. This interaction was found by Hans Burgeff of the University of Goettingen to be due to the exchange of low molecular weight substances that diffused through the substratum and atmosphere. This work constituted the first demonstration of sex hormone activity in any fungus. The elucidation of the hormonal control of sexual interaction in the Mucorales extends over 60 years and involved mycologists and biochemists from Germany, Italy, the Netherlands, the UK and the USA.

Functions of Trisporic Acid in Mucorales

Recognition of compatible sexual partners in zygomycota is based on a cooperative biosynthesis pathway of trisporic acid. Early trisporoid derivatives and trisporic acid induce swelling of two potential hyphae, hence called zygophores, and a chemical gradient of these inducer molecules results in a growth towards each other. These progametangia come in contact with each other and build a strong connection. In the next stage, septae are established to limit the developing zygospore from the vegetative mycelium and in this way the zygophores become suspensor hyphae and gametangia are formed. After dissolving of the fusion wall, cytoplasm and a high number of nuclei from both gametangia are mixed. A selectional process (unstudied) results in a reduction of nuclei and meiosis takes place (also unstudied until today). Several cell wall modifications, as well as incorporation of sporopollenin (responsible for the dark colour of spores) take place resulting in a mature zygospore.

Trisporic acid, as the endpoint of this recognition pathway, can solely be produced in presence of both compatible partners, which enzymatically produce trisporoid precursors to be further utilized by the potential sexual partner. Species specificity of these reactions is among others obtained by spatial segregation, physicochemical features of derivatives (volatility and light sensitivity), chemical modifications of trisporoids and transcriptional/posttranscriptional regulation.

Parasexualism

Trisporoids are also used in the mediation of the recognition between parasite and host. An example is the host-parasite interaction of a parasexual nature observed between *Parasitella parasitica*, a facultative mycoparasite of zygomycetes, and *Absidia glauca*.

This interaction is an example for biotrophic fusion parasitism, because genetic information is transferred into the host. Many morphological similarities in comparison to zygospore formation are seen, but the mature spore is called a sikyospore and is parasitic. During this process, gall-like structures are produced by the host *Absidia glauca*.

This, coupled with further evidence, has led to the assumption that trisporoids are not strictly species-specific, but that they might represent the general principle of mating recognition in Mucorales.

Phototropism

Light regulation has been investigated in the zygomycetes *Phycomyces blakesleeanus*, *Mucor circinelloides* and *Pilobolus crystallinus*. For example, in *Pilobolus crystallinus* light is responsible for the dispersal mechanism and the sporangiophores of *Phycomyces blakesleeanus* grow towards light. When light, particularly blue light, is involved in the regulation of fungal development, it directs the growth of fungal structures and activates metabolic pathways.

For instance, the zygomycota use light as signal to promote vegetative reproduction and growth of aerial hyphae to facilitate spore dispersal. Fungal phototropism has been investigated in detail using the fruiting body, sporangiophore, of *Phycomyces* as a model. *Phycomyces* has a complex photoreceptor system. It is able to react to different light intensities and different wavelengths. In contrast to the positive reaction to blue light, there is also a negative reaction to UV light. Reactions to red light were also observed.

Activation of Beta-Carotene Biosynthesis by Light

The two genes for the enzymes phytoene desaturase (carB) and the bifunctional phytoene synthase/carotene cyclase (carRA in *Phycomyces*, carRP in *Mucor*) are responsible for synthesis of beta-carotene. The product of the gene crgA, which was found in Mucor suppresses the carotene formation by inhibiting the accumulation of carB and carRP mRNAs.

Influence of Light in Sporulation and Sexual Development

The zygomycete *P. blakesleeanus* builds two types of sporangiophores, the macrophores and the microphores which differ in size. The formation of these sporangiophores work at different light fluences and therefore with specific photoreceptors. Light also regulates asexual sporulation. In *Mucor* the product of the crgA gene acts as an activator. In contrast, the sexual development of Phycomyces is inhibited by light because of a specialized photoreceptor system.

Gravitropism

Gravitropism is a turning or growth movement by a plant or fungus in response to gravity. It is equally widespread in both kingdoms. Statolites are required in both fungi and plants for the mechanism of gravity-sensing. The Zygomycota sporangiophores originate from specialized “basal hyphae” and pass through several distinctive developmental stages until the mature asexual spores are released. In addition to the positive phototropism, the sporangiophores are directed by a negative gravitropic response into a position suitable for spore dispersal and distribution.

Both responses are growth reactions i.e. the bending is caused by differential growth on the respective opposite flanks of the sporangiophore, and influence each other. The only model for the mechanism of the gravitropic reaction of *Phycomyces* is based on the floatability of the vacuole within the surrounding cytoplasm.

The resulting asymmetric distribution of the cytoplasm is proposed to generate increased wall growth on the lower side of horizonally placed sporangiophores as in the thicker cytoplasmic layer forming there the number of vesicles secreting cell-wall material would be higher than on the upper side. Gravitropic bending starts after approximately 15 – 30 min in horizontally placed sporangiophores and continues until after, approximately 12 – 14 hours, the sporangiophore tip has recovered its original vertical position.

Usually, the gravitropic response is weaker compared to the phototrophic one. However, in certain conditions, equilibrium could be established and the responses are comparable. In plants and fungi, phototropism and gravitropism interact in a complex manner. During continuous irradiation with unilateral light, the sporangiophore (fruiting body) of the zygomycete fungus, Phycomyces blakesleeanus reach a bending angle of photogravitropic equilibrium at which the gravitropic and phototropic stimuli balance each other.

Protein Crystals Involved in Graviperception

In *Phycomyces blakesleeanus*, wild type sporangiophores contain large, easily seen octahedral paracrystalline crystals with size up to 5×5×5 μm. Generally, they are found near the main vacuole in clusters consisting of more than ten crystals. They are often associated to the vacuolar transepts. Sedimentation with speed of about 100 μm/s can be observed when the sporangiophores are tilted. Sliding along during sedimentation or pulling at the vacuolar membranes and transepts serves as an inter-cellular signal to a probable cytoskeleton response, and that activates receptors located in the cell membrane. These receptors in turn trigger a chain of events which finally leads to the asymmetrical growth of the cell wall. Studies of the bending angle of wild type and mutant strain sporangiophore growth have shown that mutant strains that do not have crystals exhibit reduced gravitropic response.

Lipid Droplets Involved in Graviperception

Complex of apical lipid globules are also involved in graviperception. These lipids are clustered in cellular structures, complex of lipid globules, about 0.1mm below the very tip of the apex. The globules migrate to the columella when the sporangium is formed. In mature stage this complex is believed to act as a gravireceptor due to its floatability. Mutants that lack this lipid complex show greatly lowered gravitropic response.

Industrial Uses

Many species of zygomycetes can be used in important industrial processes. A resume of them is presented in the table.

Species	Product	Uses
Several Mucor and Rhizopus spp.	Lipases and proteases	Leather, detergent and medical industry (steroid transformation)
Rhizopus	Cellulases	Food production (i.e., tempeh)
R. oryzae, other Rhizopus spp.	Fumaric acid	Diverse
Rhizopus spp.	Lactic acid	Diverse
R. delemar	Biotin	Diverse
Mortierella romanniana, Mortierella vinacea and Mucor indicus	Linolenic acid	Diverse
Mortierella alpina	Arachidonic acid	Diverse
Blakeslea trispora	β-carotene	Diverse

Culture Conditions

The zygomycetes are able to grow in a wide range of environments. Most of them are mesophilic (growing at 10–40°C with an optimum 20–35°C), but some, like *Mucor miehei* or *Mucor pusillus*, are thermophilic with a minimum growth temperature of about 20°C and maximum extending up to 60°C. Others like *Mucor hiemalis* can grow at temperatures below 0°C.

Some species of the order Mucorales are able to grow under anaerobic conditions, while most of them require aerobic conditions. Furthermore, while the majority of the zygomycetes only grow at high water activities, some of them are able to grow in salt concentrations of at least 15%. Most species of *Mucor* grow rapidly on agar at room temperature filling the Petri dish in 2–3 days with their coarse aerial mycelium. When incubated in liquid culture under semi-anaerobic conditions, several species grow in yeast like state. Zygospore formation may be stimulated at higher temperatures of incubation (30–40°C).

Growth of Zygomycota in solid agar can produce low or very high fibrous colony that rapidly fills the entire Petri dish. Its color may range from pure white to shades of gray or brown. In old cultures, dark pigmented sporangia are observed. Everything depends on the species and the media used. In liquid culture, Zygomycota usually form a bland mass and do not produce spores. This is because they cannot grow aerial hyphae.

Culture Media

Zygomycetes grow well on most standard fungal culture medium such as Sabouraud dextrose agar. They can also grow on both selective and non-selective media. Minimal media, supplementary media and induction media can also be used. Most zygomycetes are sensitive to cycloheximide (actidione) and this agent should not be used in culture media.

Reproduction

A common example of a zygomycete is black bread mold (*Rhizopus stolonifer*), a member of the Mucorales. It spreads over the surface of bread and other food sources, sending hyphae inward to absorb nutrients. In its asexual phase it develops bulbous black sporangia at the tips of upright hyphae, each containing hundreds of haploid spores.

As in most zygomycetes, asexual reproduction is the most common form of reproduction. Sexual reproduction in *Rhizopus stolonifer*, as in other zygomycetes, occurs when haploid hyphae of different mating types are in close proximity to each other. Growth of the gametangia commences after gametangia come in contact, and plasmogamy, or the fusion of the cytoplasm, occurs. Karyogamy, which is the fusion of the nuclei, follows closely after. The zygosporangia are then diploid. Zygosporangia are typically thick-walled, highly resilient to environmental hardships, and metabolically inert. When conditions improve, however, they germinate to produce a sporangium or vegetative hyphae. Meiosis occurs during germination of the zygosporangium so the resulting spores or hyphae are haploid. Grows in warm and damp conditions.

Some zygomycetes disperse their spores in a more precise manner than simply allowing them to drift aimlessly on air currents. *Pilobolus*, a fungus which grows on animal dung, bends its sporangiophores towards light with the help of a light sensitive pigment (beta-carotene) and then "fires" them with an explosive squirt of high-pressure cytoplasm. Sporangia can be launched as far as 2 m, placing them far away from the dung and hopefully on vegetation which will be eaten by an herbivore, eventually to be deposited with dung elsewhere. Different mechanisms for forcible spore discharge have evolved among members of the zygomycete order Entomophthorales.

Evolution of Conidia

The evolution of the conidium from the sporangiospore is the main defining difference between zygomycetes and ascomycetes. The evolution of sporangiospores typical of zygomycetes to conidia similar to those found in ascomycetes can be modeled by a series of forms seen in zygomycetes. Many zygomycetes produce multiple sporangiospores inside a single sporangium. Some have evolved multiple small sporangiola that contain few sporangiospores. In some cases, there may be a few as three spores in each sporangiolum, and a few species have sporangiola which contain just a single spore. *Choanephora*, a zygomycete, has a sporangiolum that contains one spore with a sporangium wall that is visible at the base of the sporangium. This structure is similar to a conidium, which has two, fused cell walls, an inner spore wall and an outer sporangium wall.

4.2.2 Ascomycota (Sac Fungi)

Ascomycota is a phylum of the kingdom Fungi that, together with the Basidiomycota, forms the subkingdom Dikarya. Its members are commonly known as the sac fungi or ascomycetes. It is the largest phylum of Fungi, with over 64,000 species. The defining feature of this fungal group is the "ascus", a microscopic sexual structure in which nonmotile spores, called ascospores, are formed. However, some species of Ascomycota are asexual and thus do not form asci or ascospores. Familiar examples of sac fungi include morels, truffles, brewers› and bakers› yeast, dead man›s fingers, and cup fungi. The fungal symbionts in the majority of lichens (loosely termed «ascolichens») such as *Cladonia* belong to the Ascomycota.

Ascomycota is a monophyletic group (containing all of the descendants of a common ancestor). Previously placed in the Basidiomycota along with asexual species from other fungal taxa, asexual (or anamorphic) ascomycetes are now identified and classified based on morphological or physiological similarities to ascus-bearing taxa, and by phylogenetic analyses of DNA sequences.

Ascomycetes are of particular use to humans as sources of medicinally important compounds such as antibiotics, as well as for fermenting bread, alcoholic beverages, and cheese. Examples of ascomycetes include *Penicillium* species on cheeses and those producing antibiotics for treating bacterial infectious diseases.

Many ascomycetes are pathogens, both of animals, including humans, and of plants. Examples of ascomycetes that can cause infections in humans include *Candida albicans*, *Aspergillus niger* and several tens of species that cause skin infections. The many plant-pathogenic ascomycetes include apple scab, rice blast, the ergot fungi, black knot, and the powdery mildews. The members of the genus *Cordyceps* are entomopathogenic fungi, meaning that they parasitise and kill insects. Other entomopathogenic ascomycetes have been used successfully in biological pest control, such as *Beauveria*.

Several species of ascomycetes are biological model organisms in laboratory research. Most famously, *Neurospora crassa*, several species of yeasts, and *Aspergillus* species are used in many genetics and cell biology studies.

Sexual Reproduction in Ascomycetes

Ascomycetes are 'spore shooters'. They are fungi which produce microscopic spores inside special, elongated cells or sacs, known as 'asci', which give the group its name. Asexual reproduction is the dominant form of propagation in the Ascomycota, and is responsible for the rapid spread of these fungi into new areas. Asexual reproduction of ascomycetes is very diverse from both structural and functional points of view. The most important and general is production of conidia, but chlamydospores are also frequently produced. Furthermore, Ascomycota also reproduce asexually through budding.

Conidia Formation

Asexual reproduction may occur through vegetative reproductive spores, the conidia. The asexual, non-motile haploid spores of a fungus, which are named after the Greek word for dust (conia), are hence also known as *conidiospores*. The conidiospores commonly contain one nucleus and are products of mitotic cell divisions and thus are sometimes called *mitospores*, which are genetically identical to the mycelium from which they originate.

They are typically formed at the ends of specialized hyphae, the conidiophores. Depending on the species they may be dispersed by wind or water, or by animals. Conidiophores may simply branch off from the mycelia or they may be formed in fruiting bodies.

The hypha that creates the sporing (conidiating) tip can be very similar to the normal hyphal tip, or it can be differentiated. The most common differentiation is the formation of a bottle shaped cell called a *phialide*, from which the spores are produced. Not all of these asexual structures are a single hypha. In some groups, the conidiophores (the structures that bear the conidia) are aggregated to form a thick structure.

E.g. In the order *Moniliales,* all of them are single hyphae with the exception of the aggregations, termed as coremia or synnema. These produce structures rather like corn-stokes, with many conidia being produced in a mass from the aggregated conidiophores.

The diverse conidia and conidiophores sometimes develop in asexual sporocarps with different characteristics (e.g., acervulus, pycnidium, sporodochium). Some species of ascomycetes form their structures within plant tissue, either as parasite or saprophytes. These fungi have evolved more complex asexual sporing structures, probably influenced by the cultural conditions of plant tissue as a substrate. These structures are called the *sporodochium.*

This is a cushion of conidiophores created from a pseudoparenchymatous stroma in plant tissue. The *pycnidium* is a globose to flask-shaped parenchymatous structure, lined on its inner wall with conidiophores. The *acervulus* is a flat saucer shaped bed of conidiophores produced under a plant cuticle, which eventually erupt through the cuticle for dispersal.

Budding

Asexual reproduction process in ascomycetes also involves the budding which we clearly observe in yeast. This is termed a "blastic process". It involves the blowing out or blebbing of the hyphal tip wall. The blastic process can involve all wall layers, or there can be a new cell wall synthesized which is extruded from within the old wall.

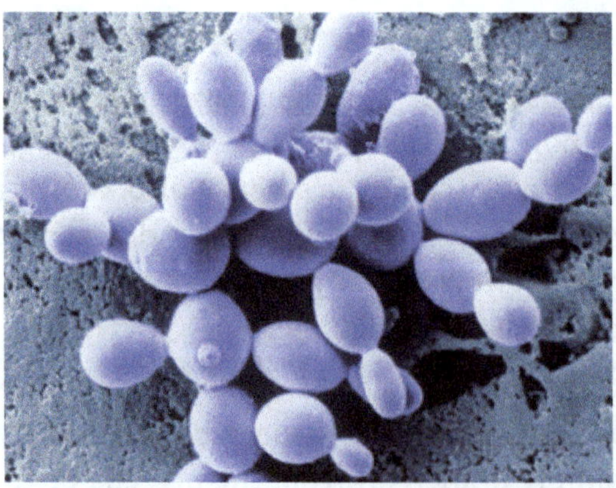

The initial events of budding can be seen as the development of a ring of chitin around the point where the bud is about to appear. This reinforces and stabilizes the cell wall. Enzymatic activity and turgor pressure act to weaken and extrude the cell wall.

New cell wall material is incorporated during this phase. Cell contents are forced into the progeny cell, and as the final phase of mitosis ends a cell plate, the point at which a new cell wall will grow inwards from, forms.

Characteristics of Ascomycetes

- Ascomycota are morphologically diverse. The group includes organisms from unicellular yeasts to complex cup fungi.
- 98% of lichens have an Ascomycota as the fungal part of the lichen.
- There are 2000 identified genera and 30,000 species of Ascomycota.
- The unifying characteristic among these diverse groups is the presence of a reproductive structure known as the *ascus*, though in some cases it has a reduced role in the life cycle.
- Many ascomycetes are of commercial importance. Some play a beneficial role, such as the yeasts used in baking, brewing, and wine fermentation, plus truffles and morels, which are held as gourmet delicacies.
- Many of them cause tree diseases, such as Dutch elm disease and apple blights.
- Some of the plant pathogenic ascomycetes are apple scab, rice blast, the ergot fungi, black knot, and the powdery mildews.
- The yeasts are used to produce alcoholic beverages and breads. The mold *Penicillium* is used to produce the antibiotic penicillin.
- Almost half of all members of the phylum Ascomycota form *symbiotic* associations with algae to form lichens.
- Others, such as morels (a highly prized edible fungi), form important *mycorrhizal* relationships with plants, thereby providing enhanced water and nutrient uptake and, in some cases, protection from insects.
- Most ascomycetes are terrestrial or parasitic. However, some have adapted to marine or freshwater environments. As of 2015, there were 805 marine fungi in the Ascomycota, distributed among 352 genera.
- The cell walls of the hyphae are variably composed of chitin and β-glucans, just as in Basidiomycota. However, these fibers are set in a matrix of glycoprotein containing the sugars galactose and mannose.

- The mycelium of ascomycetes is usually made up of septate hyphae. However, there is not necessarily any fixed number of nuclei in each of the divisions.
- The septal walls have septal pores which provide cytoplasmic continuity throughout the individual hyphae. Under appropriate conditions, nuclei may also migrate between septal compartments through the septal pores.
- A unique character of the Ascomycota (but not present in all ascomycetes) is the presence of Woronin bodies on each side of the septa separating the hyphal segments which control the septal pores. If an adjoining hypha is ruptured, the Woronin bodies block the pores to prevent loss of cytoplasm into the ruptured compartment. The Woronin bodies are spherical, hexagonal, or rectangular membrane bound structures with a crystalline protein matrix.

Modern Classification

There are three subphyla that are described and accepted:

- The *Pezizomycotina* are the largest subphylum and contains all ascomycetes that produce ascocarps (fruiting bodies), except for one genus, *Neolecta*, in the Taphrinomycotina. It is roughly equivalent to the previous taxon, *Euascomycetes*. The Pezizomycotina includes most macroscopic "ascos" such as truffles, ergot, ascolichens, cup fungi (discomycetes), pyrenomycetes, lorchels, and caterpillar fungus. It also contains microscopic fungi such as powdery mildews, dermatophytic fungi, and Laboulbeniales.
- The *Saccharomycotina* comprise most of the "true" yeasts, such as baker's yeast and *Candida*, which are single-celled (unicellular) fungi, which reproduce vegetatively by budding. Most of these species were previously classified in a taxon called *Hemiascomycetes*.
- The *Taphrinomycotina* include a disparate and basal group within the Ascomycota that was recognized following molecular (DNA) analyses. The taxon was originally named *Archiascomycetes* (or *Archaeascomycetes*). It includes hyphal fungi (*Neolecta*, *Taphrina*, *Archaeorhizomyces*), fission yeasts (*Schizosaccharomyces*), and the mammalian lung parasite *Pneumocystis*.

Outdated Taxon Names

Several outdated taxon names—based on morphological features—are still occasionally used for species of the Ascomycota. These include the following sexual (teleomorphic) groups, defined by the structures of their sexual fruiting bodies: the Discomycetes, which included all species forming apothecia; the Pyrenomycetes, which included all sac fungi that formed perithecia or pseudothecia, or any structure resembling these morphological structures; and the Plectomycetes, which included those species that form cleistothecia. Hemiascomycetes included the yeasts and yeast-like fungi that have now been placed into the Saccharomycotina or Taphrinomycotina, while the Euascomycetes included the remaining species of the Ascomycota, which are now in the Pezizomycotina, and the Neolecta, which are in the Taphrinomycotina.

Some ascomycetes do not reproduce sexually or are not known to produce asci and are therefore anamorphic species. Those anamorphs that produce conidia (mitospores) were previously described as mitosporic Ascomycota. Some taxonomists placed this group into a separate artificial phylum, the Deuteromycota (or "Fungi Imperfecti"). Where recent molecular analyses have identified close relationships with ascus-bearing taxa, anamorphic species have been grouped into the Ascomycota, despite the absence of the defining ascus. Sexual and asexual isolates of the same species commonly carry different binomial species names, as, for example, *Aspergillus nidulans* and *Emericella nidulans*, for asexual and sexual isolates, respectively, of the same species.

Species of the Deuteromycota were classified as Coelomycetes if they produced their conidia in minute flask- or saucer-shaped conidiomata, known technically as *pycnidia* and *acervuli*. The Hyphomycetes were those species where the conidiophores (*i.e.*, the hyphal structures that carry conidia-forming cells at the end) are free or loosely organized. They are mostly isolated but sometimes also appear as bundles of cells aligned in parallel (described as *synnematal*) or as cushion-shaped masses (described as *sporodochial*).

Morphology

Most species grow as filamentous, microscopic structures called hyphae or as budding single cells (yeasts). Many interconnected hyphae form a thallus usually referred to as the mycelium, which—when visible to the naked eye (macroscopic)—is commonly called mold. During sexual reproduction, many Ascomycota typically produce large numbers of asci. The ascus is often contained in a multicellular, occasionally readily visible fruiting structure, the ascocarp (also called an *ascoma*). Ascocarps come in a very large variety of shapes: cup-shaped, club-shaped, potato-like, spongy, seed-like, oozing and pimple-like, coral-like, nit-like, golf-ball-shaped, perforated tennis ball-like, cushion-shaped, plated and feathered in miniature (Laboulbeniales), microscopic classic Greek shield-shaped, stalked or sessile. They can appear solitary or clustered. Their texture can likewise be very variable, including fleshy, like charcoal (carbonaceous), leathery, rubbery, gelatinous, slimy, powdery, or cob-web-like. Ascocarps come in multiple colors such as red, orange, yellow, brown, black, or, more rarely, green or blue. Some ascomyceous fungi, such as *Saccharomyces cerevisiae*, grow as single-celled yeasts, which—during sexual reproduction—develop into an ascus, and do not form fruiting bodies.

In lichenized species, the thallus of the fungus defines the shape of the symbiotic colony. Some dimorphic species, such as *Candida albicans*, can switch between growth as single cells and as filamentous, multicellular hyphae. Other species are pleomorphic, exhibiting asexual (anamorphic) as well as a sexual (teleomorphic) growth forms.

Except for lichens, the non-reproductive (vegetative) mycelium of most ascomycetes is usually inconspicuous because it is commonly embedded in the substrate, such as soil, or grows on or inside a living host, and only the ascoma may be seen when fruiting. Pigmentation, such as melanin in hyphal walls, along with prolific growth on surfaces can result in visible mold colonies; examples include *Cladosporium* species, which form black spots on bathroom caulking and other moist areas. Many ascomycetes cause food spoilage, and, therefore, the pellicles or moldy layers that develop on jams, juices, and other foods are the mycelia of these species or occasionally Mucoromycotina and almost never Basidiomycota. Sooty molds that develop on plants, especially in the tropics are the thalli of many species.

Large masses of yeast cells, asci or ascus-like cells, or conidia can also form macroscopic structures. For example. *Pneumocystis* species can colonize lung cavities (visible in x-rays), causing a form of pneumonia. Asci of *Ascosphaera* fill honey bee larvae and pupae causing mummification with a chalk-like appearance, hence the name "chalkbrood". Yeasts for small colonies in vitro and in vivo, and excessive growth of *Candida* species in the mouth or vagina causes "thrush", a form of candidiasis.

The cell walls of the ascomycetes almost always contain chitin and β-glucans, and divisions within the hyphae, called "septa", are the internal boundaries of individual cells (or compartments). The cell wall and septa give stability and rigidity to the hyphae and may prevent loss of cytoplasm in case of local damage to cell wall and cell membrane. The septa commonly have a small opening in the center, which functions as a cytoplasmic connection between adjacent cells, also sometimes allowing cell-to-cell movement of nuclei within a hypha. Vegetative hyphae of most ascomycetes contain only one nucleus per cell (*uninucleate* hyphae), but multinucleate cells—especially in the apical regions of growing hyphae—can also be present.

Metabolism

In common with other fungal phyla, the Ascomycota are heterotrophic organisms that require organic compounds as energy sources. These are obtained by feeding on a variety of organic substrates including dead matter, foodstuffs, or as symbionts in or on other living organisms. To obtain these nutrients from their surroundings, ascomycetous fungi secrete powerful digestive enzymes that break down organic substances into smaller molecules, which are then taken up into the cell. Many species live on dead plant material such as leaves, twigs, or logs. Several species colonize plants, animals, or other fungi as parasites or mutualistic symbionts and derive all their metabolic energy in form of nutrients from the tissues of their hosts.

Owing to their long evolutionary history, the Ascomycota have evolved the capacity to break down almost every organic substance. Unlike most organisms, they are able to use their own enzymes to digest plant biopolymers such as cellulose or lignin. Collagen, an abundant structural protein in animals, and keratin—a protein that forms hair and nails, can also serve as food sources. Unusual examples include *Aureobasidium pullulans*, which feeds on wall paint, and the kerosene fungus *Amorphotheca resinae*, which feeds on aircraft fuel (causing occasional problems for the airline industry), and may sometimes block fuel pipes. Other species can resist high osmotic stress and grow, for example, on salted fish, and a few ascomycetes are aquatic.

The Ascomycota is characterized by a high degree of specialization; for instance, certain species of Laboulbeniales attack only one particular leg of one particular insect species. Many Ascomycota engage in symbiotic relationships such as in lichens—symbiotic associations with green algae or cyanobacteria—in which the fungal symbiont directly obtains products of photosynthesis. In common with many basidiomycetes and Glomeromycota, some ascomycetes form symbioses with plants by colonizing the roots to form mycorrhizal associations. The Ascomycota also represents several carnivorous fungi, which have developed hyphal traps to capture small protists such as amoebae, as well as roundworms (*Nematoda*), rotifers, tardigrades, and small arthropods such as springtails (*Collembola*).

Distribution and Living Environment

The Ascomycota are represented in all land ecosystems worldwide, occurring on all continents including Antarctica. Spores and hyphal fragments are dispersed through the atmosphere and freshwater environments, as well as ocean beaches and tidal zones.

The distribution of species is variable; while some are found on all continents, others, as for example the white truffle *Tuber magnatum*, only occur in isolated locations in Italy and Eastern Europe. The distribution of plant-parasitic species is often restricted by host distributions; for example, *Cyttaria* is only found on *Nothofagus* (Southern Beech) in the Southern Hemisphere.

Reproduction

Asexual Reproduction

Asexual reproduction is the dominant form of propagation in the Ascomycota, and is responsible for the rapid spread of these fungi into new areas. It occurs through vegetative reproductive spores, the conidia.

The conidiospores commonly contain one nucleus and are products of mitotic cell divisions and thus are sometimes called mitospores, which are genetically identical to the mycelium from which they originate. They are typically formed at the ends of specialized hyphae, the *conidiophores*. Depending on the species they may be dispersed by wind or water, or by animals.

Asexual Spores

Different types of asexual spores can be identified by colour, shape, and how they are released as individual spores. Spore types can be used as taxonomic characters in the classification within the Ascomycota. The most frequent types are the single-celled spores, which are designated *amerospores*. If the spore is divided into two by a cross-wall (septum), it is called a *didymospore*.

When there are two or more cross-walls, the classification depends on spore shape. If the septae are *transversal*, like the rungs of a ladder, it is a *phragmospore*, and if they possess a net-like structure it is a *dictyospore*. In *staurospores* ray-like arms radiate from a central body; in others (*helicospores*) the entire spore is wound up in a spiral like a spring. Very long worm-like spores with a length-to-diameter ratio of more than 15:1, are called *scolecospores*.

Conidiogenesis and Dehiscence

Important characteristics of the anamorphs of the Ascomycota are *conidiogenesis*, which includes spore formation and dehiscence (separation from the parent structure).

Conidiogenesis corresponds to Embryology in animals and plants and can be divided into two fundamental forms of development: *blastic* conidiogenesis, where the spore is already evident before it separates from the conidiogenic hypha, and *thallic* conidiogenesis, during which a cross-wall forms and the newly created cell develops into a spore.

The spores may or may not be generated in a large-scale specialized structure that helps to spread them.

Basic types can be further classified as follows:

- **Blastic-acropetal** (repeated budding at the tip of the conidiogenic hypha, so that a chain of spores is formed with the youngest spores at the tip).

- **Blastic-synchronous** (simultaneous spore formation from a central cell, sometimes with secondary acropetal chains forming from the initial spores).
- **Blastic-sympodial** (repeated sideways spore formation from behind the leading spore, so that the oldest spore is at the main tip).
- **Blastic-annellidic** (each spore separates and leaves a ring-shaped scar inside the scar left by the previous spore).
- **Blastic-phialidic** (the spores arise and are ejected from the open ends of special conidiogenic cells called phialides, which remain constant in length).
- **Basauxic** (where a chain of conidia, in successively younger stages of development, is emitted from the mother cell).
- **Blastic-retrogressive** (spores separate by formation of crosswalls near the tip of the conidiogenic hypha, which thus becomes progressively shorter).
- **Thallic-arthric** (double cell walls split the conidiogenic hypha into cells that develop into short, cylindrical spores called *arthroconidia*; sometimes every second cell dies off, leaving the arthroconidia free).
- **Thallic-solitary** (a large bulging cell separates from the conidiogenic hypha, forms internal walls, and develops to a *phragmospore*).

Sometimes the conidia are produced in structures visible to the naked eye, which help to distribute the spores. These structures are called "conidiomata" (singular: conidioma), and may take the form of *pycnidia* (which are flask-shaped and arise in the fungal tissue) or *acervuli* (which are cushion-shaped and arise in host tissue).

Dehiscence happens in two ways. In *schizolytic* dehiscence, a double-dividing wall with a central lamella (layer) forms between the cells; the central layer then breaks down thereby releasing the spores. In *rhexolytic* dehiscence, the cell wall that joins the spores on the outside degenerates and releases the conidia.

Heterokaryosis and Parasexuality

Several Ascomycota species are not known to have a sexual cycle. Such asexual species may be able to undergo genetic recombination between individuals by processes involving *heterokaryosis* and *parasexual* events.

Parasexuality refers to the process of heterokaryosis, caused by merging of two hyphae belonging to different individuals, by a process called *anastomosis*, followed by a series of events resulting in genetically different cell nuclei in the mycelium. The merging of nuclei is not followed by meiotic events, such as gamete formation and results in an increased number of chromosomes per nuclei. *Mitotic crossover* may enable recombination, i.e., an exchange of genetic material between homologous chromosomes. The chromosome number may then be restored to its haploid state by nuclear division, with each daughter nuclei being genetically different from the original parent nuclei. Alternatively, nuclei may lose some chromosomes, resulting in aneuploid cells. *Candida albicans* (class Saccharomycetes) is an example of a fungus that has a parasexual cycle.

Sexual Reproduction

Sexual reproduction in the Ascomycota leads to the formation of the *ascus*, the structure that defines this fungal group and distinguishes it from other fungal phyla. The ascus is a tube-shaped vessel, a *meiosporangium*, which contains the sexual spores produced by meiosis and which are called *ascospores.*

Apart from a few exceptions, such as *Candida albicans*, most ascomycetes are haploid, i.e., they contain one set of chromosomes per nucleus. During sexual reproduction there is a diploid phase, which commonly is very short, and meiosis restores the haploid state. The sexual cycle of one well-studied representative species of Ascomycota is described in greater detail in *Neurospora crassa*. Also, the adaptive basis for the maintenance of sexual reproduction in the Ascomycota fungi was reviewed by Wallen and Perlin. They concluded that the most plausible reason for the maintenance of this capability is the benefit of repairing DNA damage by using recombination that occurs during meiosis. DNA damage can be caused by a variety of stresses such as nutrient limitation.

Formation of Sexual Spores

The sexual part of the life cycle commences when two hyphal structures mate. In the case of *homothallic* species, mating is enabled between hyphae of the same fungal clone, whereas in *heterothallic* species, the two hyphae must originate from fungal clones that differ genetically, i.e., those that are of a different mating type. Mating types are typical of the fungi and correspond roughly to the sexes in plants and animals; however one species may have more than two mating types, resulting in sometimes complex vegetative incompatibility systems. The adaptive function of mating type is discussed in *Neurospora crassa*.

Gametangia are sexual structures formed from hyphae, and are the generative cells. A very fine hypha, called trichogyne emerges from one gametangium, the *ascogonium*, and merges with a gametangium (the *antheridium*) of the other fungal isolate. The nuclei in the antheridium then migrate into the ascogonium, and plasmogamy—the mixing of the cytoplasm—occurs. Unlike in animals and plants, plasmogamy is not immediately followed by the merging of the nuclei (called *karyogamy*). Instead, the nuclei from the two hyphae form pairs, initiating the *dikaryophase* of the sexual cycle, during which time the pairs of nuclei synchronously divide. Fusion of the paired nuclei leads to mixing of the genetic material and recombination and is followed by meiosis. A similar sexual cycle is present in the red algae (Rhodophyta). A discarded hypothesis held that a second karyogamy event occurred in the ascogonium prior to ascogeny, resulting in a tetraploid nucleus which divided into four diploid nuclei by meiosis and then into eight haploid nuclei by a supposed process called brachymeiosis, but this hypothesis was disproven in the 1950s.

From the fertilized ascogonium, *dinucleate* hyphae emerge in which each cell contains two nuclei. These hyphae are called *ascogenous* or fertile hyphae. They are supported by the vegetative mycelium containing uni– (or mono–) nucleate hyphae, which are sterile. The mycelium containing both sterile and fertile hyphae may grow into fruiting body, the *ascocarp*, which may contain millions of fertile hyphae.

An ascocarp is the fruiting body of the sexual phase in Ascomycota. There are five morphologically different types of ascocarp, namely:

- **Naked Asci:** These occur in simple ascomycetes; asci are produced on the organism's surface.
- **Perithecia:** Asci are in flask-shaped ascoma (perithecium) with a pore (ostiole) at the top.
- **Cleistothecia:** The ascocarp (a cleistothecium) is spherical and closed.
- **Apothecia:** The asci are in a bowl shaped ascoma (apothecium). These are sometimes called the "cup fungi".
- **Pseudothecia:** Asci with two layers, produced in pseudothecia that look like perithecia. The ascospores are arranged irregularly.

The sexual structures are formed in the fruiting layer of the ascocarp, the hymenium. At one end of ascogenous hyphae, characteristic U-shaped hooks develop, which curve back opposite to the growth direction of the hyphae. The two nuclei contained in the apical part of each hypha divide in such a way that the threads of their mitotic spindles run parallel, creating two pairs of genetically different nuclei. One daughter nucleus migrates close to the hook, while the other daughter nucleus locates to the basal part of the hypha. The formation of two parallel cross-walls then divides the hypha into three sections: one at the hook with one nucleus, one at the basal of the original hypha that contains one nucleus, and one that separates the U-shaped part, which contains the other two nuclei.

Fusion of the nuclei (karyogamy) takes place in the U-shaped cells in the hymenium, and results in the formation of a diploid zygote. The zygote grows into the ascus, an elongated tube-shaped or cylinder-shaped capsule. Meiosis then gives rise to four haploid nuclei, usually followed by a further mitotic division that results in eight nuclei in each ascus. The nuclei along with some cytoplasma become enclosed within membranes and a cell wall to give rise to ascospores that are aligned inside the ascus like peas in a pod.

Upon opening of the ascus, ascospores may be dispersed by the wind, while in some cases the spores are forcibly ejected form the ascus; certain species have evolved spore cannons, which can eject ascospores up to 30 cm. away. When the spores reach a suitable substrate, they germinate, form new hyphae, which restarts the fungal life cycle.

The form of the ascus is important for classification and is divided into four basic types: unitunicate-operculate, unitunicate-inoperculate, bitunicate, or prototunicate.

Ecology

The Ascomycota fulfil a central role in most land-based ecosystems. They are important decomposers, breaking down organic materials, such as dead leaves and animals, and helping the detritivores (animals that feed on decomposing material) to obtain their nutrients. Ascomycetes, along with other fungi, can break down large molecules such as cellulose or lignin, and thus have important roles in nutrient cycling such as the carbon cycle. The fruiting bodies of the Ascomycota provide food for many animals ranging from insects and slugs and snails (*Gastropoda*) to rodents and larger mammals such as deer and wild boars. Many ascomycetes also form symbiotic relationships with other organisms, including plants and animals.

Lichens

Probably since early in their evolutionary history, the Ascomycota have formed symbiotic associations with green algae (*Chlorophyta*), and other types of algae and cyanobacteria. These mutualistic associations are commonly known as lichens, and can grow and persist in terrestrial regions of the earth that are inhospitable to other organisms and characterized by extremes in temperature and humidity, including the Arctic, the Antarctic, deserts, and mountaintops. While the photoautotrophic algal partner generates metabolic energy through photosynthesis, the fungus offers a stable, supportive matrix and protects cells from radiation and dehydration. Around 42% of the Ascomycota (about 18,000 species) form lichens, and almost all the fungal partners of lichens belong to the Ascomycota.

Mycorrhizal Fungi and Endophytes

Members of the Ascomycota form two important types of relationship with plants: as mycorrhizal fungi and as endophytes. Mycorrhiza are symbiotic associations of fungi with the root systems of the plants, which can be of vital importance for growth and persistence for the plant. The fine mycelial network of the fungus enables the increased uptake of mineral salts that occur at low levels in the soil. In return, the plant provides the fungus with metabolic energy in the form of photosynthetic products.

Endophytic fungi live inside plants, and those that form mutualistic or commensal associations with their host, do not damage their hosts. The exact nature of the relationship between endophytic fungus and host depends on the species involved, and in some cases fungal colonization of plants can bestow a higher resistance against insects, roundworms (nematodes), and bacteria; in the case of grass endophytes the fungal symbiont produces poisonous alkaloids, which can affect the health of plant-eating (herbivorous) mammals and deter or kill insect herbivores.

Symbiotic Relationships with Animals

Several ascomycetes of the genus *Xylaria* colonize the nests of leafcutter ants and other fungus-growing ants of the tribe Attini, and the fungal gardens of termites (Isoptera). Since they do not generate fruiting bodies until the insects have left the nests, it is suspected that, as confirmed in several cases of Basidiomycota species, they may be cultivated.

Bark beetles (family Scolytidae) are important symbiotic partners of ascomycetes. The female beetles transport fungal spores to new hosts in characteristic tucks in their skin, the *mycetangia*. The beetle tunnels into the wood and into large chambers in which they lay their eggs. Spores released from the mycetangia germinate into hyphae, which can break down the wood. The beetle larvae then feed on the fungal mycelium, and, on reaching maturity, carry new spores with them to renew the cycle of infection. A well-known example of this is Dutch elm disease, caused by *Ophiostoma ulmi*, which is carried by the European elm bark beetle, *Scolytus multistriatus*.

Plant Disease Interactions

One of their most harmful roles is as the agent of many plant diseases. For instance:

- Dutch elm disease, caused by the closely related species *Ophiostoma ulmi* and *Ophiostoma novo-ulmi*, has led to the death of many elms in Europe and North America.
- The originally Asian *Cryphonectria parasitica* is responsible for attacking Sweet Chestnuts (*Castanea sativa*), and virtually eliminated the once-widespread American Chestnut (*Castanea dentata*).
- A disease of maize (*Zea mays*), which is especially prevalent in North America, is brought about by *Cochliobolus heterostrophus*.
- *Taphrina deformans* causes leaf curl of peach.

- *Uncinula necator* is responsible for the disease powdery mildew, which attacks grapevines.
- Species of *Monilinia* cause brown rot of stone fruit such as peaches (*Prunus persica*) and sour cherries (*Prunus ceranus*).
- Members of the Ascomycota such as *Stachybotrys chartarum* are responsible for fading of woolen textiles, which is a common problem especially in the tropics.
- Blue-green, red and brown molds attack and spoil foodstuffs – for instance *Penicillium italicum* rots oranges.
- Cereals infected with *Fusarium graminearum* contain mycotoxins like deoxynivalenol (DON), which causes Fusarium ear blight and skin and mucous membrane lesions when eaten by pigs.

Human Disease Interactions

- **Aspergillus fumigatus**, the most common cause of fungal infection in the lungs of immune-compromised patients often resulting in death. Also the most frequent cause of Allergic bronchopulmonary aspergillosis, which often occurs in patients with Cystic fibrosis as well as Asthma.
- **Candida albicans**, a yeast that attacks the mucous membranes, can cause an infection of the mouth or vagina called thrush or candidiasis, and is also blamed for "yeast allergies".
- Fungi like **Epidermophyton** cause skin infections but are not very dangerous for people with healthy immune systems. However, if the immune system is damaged they can be life-threatening; for instance, *Pneumocystis jirovecii* is responsible for severe lung infections that occur in AIDS patients.
- **Ergot (Claviceps purpurea)** is a direct menace to humans when it attacks wheat or rye and produces highly poisonous alkaloids, causing ergotism if consumed. Symptoms include hallucinations, stomach cramps, and a burning sensation in the limbs ("Saint Anthony's Fire").
- **Aspergillus flavus**, which grows on peanuts and other hosts, generates aflatoxin, which damages the liver and is highly carcinogenic.
- **Histoplasma capsulatum** causes histoplasmosis, which affects immunocompromised patients.
- **Blastomyces dermatitidis** is the causal agent of blastomycosis, an invasive and often serious fungal infection found occasionally in humans and other animals in regions where the fungus is endemic.

- **Paracoccidioides brasiliensis and Paracoccidioides lutzii** are the causal agents of paracoccidioidomycosis.
- **Coccidioides immitis** and *Coccidioides posadasii* are the causative agent of coccidioidomycosis (valley fever).
- **Talaromyces marneffei**, formerly called *Penicillium marneffei* causes talaromycosis.

Beneficial Effects for Humans

On the other hand, ascus fungi have brought some significant benefits to humanity.

- The most famous case may be that of the mold *Penicillium chrysogenum* (formerly *Penicillium notatum*), which, probably to attack competing bacteria, produces an antibiotic that, under the name of penicillin, triggered a revolution in the treatment of bacterial infectious diseases in the 20th century.
- The medical importance of *Tolypocladium niveum* as an immunosuppressor can hardly be exaggerated. It excretes Ciclosporin, which, as well as being given during Organ transplantation to prevent rejection, is also prescribed for auto-immune diseases such as multiple sclerosis. However, there is some doubt over the long-term side effects of the treatment.
- Some ascomycete fungi can be easily altered through genetic engineering procedures. They can then produce useful proteins such as insulin, human growth hormone, or TPa, which is employed to dissolve blood clots.
- Several species are common model organisms in biology, including *Saccharomyces cerevisiae*, *Schizosaccharomyces pombe*, and *Neurospora crassa*. The genomes of some ascomycete fungi have been fully sequenced.
- Baker's Yeast (*Saccharomyces cerevisiae*) is used to make bread, beer and wine, during which process sugars such as glucose or sucrose are fermented to make ethanol and carbon dioxide. Bakers use the yeast for carbon dioxide production, causing the bread to rise, with the ethanol boiling off during cooking. Most vintners use it for ethanol production, releasing carbon dioxide into the atmosphere during fermentation. Brewers and traditional producers of sparkling wine use both, with a primary fermentation for the alcohol and a secondary one to produce the carbon dioxide bubbles that provide the drinks with a "sparkling" texture in the case of wine and the desirable foam in the case of beer.

- Enzymes of *Penicillium camemberti* play a role in the manufacture of the cheeses Camembert and Brie, while those of *Penicillium roqueforti* do the same for Gorgonzola, Roquefort and Stilton.
- In Asia, *Aspergillus oryzae* is added to a pulp of soaked soya beans to make soy sauce and is used to break down starch in rice and other grains into simple sugars for fermentation into East Asian alcoholic beverages such as huangjiu and sake.
- Finally, some members of the Ascomycota are choice edibles; morels (*Morchella spp.*), truffles (*Tuber spp.*), and lobster mushroom (*Hypomyces lactifluorum*) are some of the most sought-after fungal delicacies.
- Cordyceps militaris is known for its numerous medicinal benefits, including supporting the immune system, reducing inflammation, providing antioxidant effects, enhancing metabolic health, improving athletic performance, and promoting respiratory health. It contains bioactive compounds such as cordycepin, cordycepic acid, adenosine, and polysaccharides, beta-glucans, and ergosterol.

4.2.3 Basidiomycota (Club Fungi)

Basidiomycota is one of two large divisions that, together with the Ascomycota, constitute the subkingdom Dikarya (often referred to as the "higher fungi") within the kingdom Fungi. Members are known as basidiomycetes. More specifically, Basidiomycota includes these groups: agarics, puffballs, stinkhorns, bracket fungi, other polypores, jelly fungi, boletes, chanterelles, earth stars, smuts, bunts, rusts, mirror yeasts, and *Cryptococcus*, the human pathogenic yeast.

Basidiomycota are filamentous fungi composed of hyphae (except for basidiomycota-yeast) and reproduce sexually via the formation of specialized club-shaped end cells called basidia that normally bear external meiospores (usually four). These specialized spores are called basidiospores. However, some Basidiomycota are obligate asexual reproducers. Basidiomycota that reproduce asexually can typically be recognized as members of this division by gross similarity to others, by the formation of a distinctive anatomical feature (the clamp connection), cell wall components, and definitively by phylogenetic molecular analysis of DNA sequence data.

Classification

A 2007 classification, adopted by a coalition of 67 mycologists recognized three subphyla (Pucciniomycotina, Ustilaginomycotina, Agaricomycotina) and two other class level taxa (Wallemiomycetes, Entorrhizomycetes) outside of these, among the Basidiomycota. As now classified, the subphyla join and also cut across various obsolete taxonomic groups previously commonly used to describe Basidiomycota.

According to a 2008 estimate, Basidiomycota comprise three subphyla (including six unassigned classes) 16 classes, 52 orders, 177 families, 1,589 genera, and 31,515 species.

Traditionally, the Basidiomycota were divided into two classes, now obsolete:

- Homobasidiomycetes (alternatively called holobasidiomycetes), including true mushrooms.
- Heterobasidiomycetes, including the jelly, rust and smut fungi.

Nonetheless these former concepts continue to be used as two types of growth habit groupings, the "mushrooms" (e.g. *Schizophyllum commune*) and the non-mushrooms (e.g. *Mycosarcoma maydis*).

Agaricomycotina

The Agaricomycotina include what had previously been called the Hymenomycetes (an obsolete morphological based class of Basidiomycota that formed hymenial layers on their fruitbodies), the Gasteromycetes (another obsolete class that included species mostly lacking hymenia and mostly forming spores in enclosed fruitbodies), as well as most of the jelly fungi. This sub-phyla also includes the "classic" mushrooms, polypores, corals, chanterelles, crusts, puffballs and stinkhorns. The three classes in the Agaricomycotina are the Agaricomycetes, the Dacrymycetes, and the Tremellomycetes. The class Wallemiomycetes is not yet placed in a subdivision, but recent genomic evidence suggests that it is a sister group of Agaricomycotina.

Pucciniomycotina

The Pucciniomycotina include the rust fungi, the insect parasitic/symbiotic genus *Septobasidium*, a former group of smut fungi (in the Microbotryomycetes, which includes mirror yeasts), and a mixture of odd, infrequently seen, or seldom recognized fungi, often parasitic on plants. The eight classes in the Pucciniomycotina are Agaricostilbomycetes, Atractiellomycetes, Classiculomycetes, Cryptomycocolacomycetes, Cystobasidiomycetes, Microbotryomycetes, Mixiomycetes, and Pucciniomycetes.

Ustilaginomycotina

The Ustilaginomycotina are most (but not all) of the former smut fungi and the Exobasidiales. The classes of the Ustilaginomycotina are the Exobasidiomycetes, the Entorrhizomycetes, and the Ustilaginomycetes.

Typical Life Cycle

Unlike animals and plants which have readily recognizable male and female counterparts, Basidiomycota (except for the Rust (Pucciniales)) tend to have mutually indistinguishable, compatible haploids which are usually mycelia being composed of filamentous hyphae. Typically haploid Basidiomycota mycelia fuse via plasmogamy and then the compatible nuclei migrate into each other's mycelia and pair up with the resident nuclei. Karyogamy is delayed, so that the compatible nuclei remain in pairs, called a dikaryon.

The hyphae are then said to be dikaryotic. Conversely, the haploid mycelia are called monokaryons. Often, the dikaryotic mycelium is more vigorous than the individual monokaryotic mycelia, and proceeds to take over the substrate in which they are growing. The dikaryons can be long-lived, lasting years, decades, or centuries. The monokaryons are *neither* male nor female. They have either a *bipolar* (*unifactorial*) or a *tetrapolar* (*bifactorial*) mating system.

This results in the fact that following meiosis, the resulting haploid basidiospores and resultant monokaryons, have nuclei that are compatible with 50% (if bipolar) or 25% (if tetrapolar) of their sister basidiospores (and their resultant monokaryons) because the mating genes must differ for them to be compatible. However, there are sometimes more than two possible alleles for a given locus, and in such species, depending on the specifics, over 90% of monokaryons could be compatible with each other.

The maintenance of the dikaryotic status in dikaryons in many Basidiomycota is facilitated by the formation of clamp connections that physically appear to help coordinate and re-establish pairs of compatible nuclei following synchronous mitotic nuclear divisions. Variations are frequent and multiple. In a typical Basidiomycota lifecycle the long lasting dikaryons periodically (seasonally or occasionally) produce basidia, the specialized usually club-shaped end cells, in which a pair of compatible nuclei fuse (karyogamy) to form a diploid cell. Meiosis follows shortly with the production of 4 haploid nuclei that migrate into 4 external, usually apical basidiospores. Variations occur, however.

Typically the basidiospores are ballistic, hence they are sometimes also called ballistospores. In most species, the basidiospores disperse and each can start a new haploid mycelium, continuing the lifecycle. Basidia are microscopic but they are often produced on or in multicelled large fructifications called basidiocarps or basidiomes, or fruitbodies, variously called mushrooms, puffballs, etc. Ballistic basidiospores are formed on sterigmata which are tapered spine-like projections on basidia, and are typically curved, like the horns of a bull. In some Basidiomycota the spores are not ballistic, and the sterigmata may be straight, reduced to stubs, or absent. The basidiospores of these non-ballistosporic basidia may either bud off, or be released via dissolution or disintegration of the basidia.

In summary, meiosis takes place in a diploid basidium. Each one of the four haploid nuclei migrates into its own basidiospore. The basidiospores are ballistically discharged and start new haploid mycelia called monokaryons. There are no males or females, rather there are compatible thalli with multiple compatibility factors. Plasmogamy between compatible individuals leads to delayed karyogamy leading to establishment of a dikaryon. The dikaryon is long lasting but ultimately gives rise to either fruitbodies with basidia or directly to basidia without fruitbodies. The paired dikaryon in the basidium fuse (i.e., karyogamy takes place). The diploid basidium begins the cycle again.

Meiosis

Coprinopsis cinerea is a basidiomycete mushroom. It is particularly suited to the study of meiosis because meiosis progresses synchronously in about 10 million cells within the mushroom cap, and the meiotic prophase stage is prolonged. Burns et al. studied the expression of genes involved in the 15-hour meiotic process, and found that the pattern of gene expression of *C. cinerea* was similar to two other fungal species, the yeasts *Saccharomyces cerevisiae* and *Schizosaccharomyces pombe*. These similarities in the patterns of expression led to the conclusion that the core expression program of meiosis has been conserved in these fungi for over half a billion years of evolution since these species diverged.

Cryptococcus neoformans and *Mycosarcoma maydis* are examples of pathogenic basidiomycota. Such pathogens must be able to overcome the oxidative defenses of their respective hosts in order to produce a successful infection. The ability to undergo meiosis may provide a survival benefit for these fungi by promoting successful infection. A characteristic central feature of meiosis is recombination between homologous chromosomes. This process is associated with repair of DNA damage, particularly double-strand breaks. The ability of *C. neoformans* and *M. maydis* to undergo meiosis may contribute to their virulence by repairing the oxidative DNA damage caused by their host's release of reactive oxygen species.

Variations in Lifecycles

Many variations occur: some variations are self-compatible and spontaneously form dikaryons without a separate compatible thallus being involved. These fungi are said to be homothallic, versus the normal heterothallic species with mating types. Others are secondarily homothallic, in that two compatible nuclei following meiosis migrate into each basidiospore, which is then dispersed as a pre-existing dikaryon.

Often such species form only two spores per basidium, but that too varies. Following meiosis, mitotic divisions can occur in the basidium. Multiple numbers of basidiospores can result, including odd numbers via degeneration of nuclei, or pairing up of nuclei, or lack of migration of nuclei. For example, the chanterelle genus *Craterellus* often has six-spored basidia, while some corticioid *Sistotrema* species can have two-, four-, six-, or eight-spored basidia, and the cultivated button mushroom, *Agaricus bisporus.* can have one-, two-, three- or four-spored basidia under some circumstances. Occasionally, monokaryons of some taxa can form morphologically fully formed basidiomes and anatomically correct basidia and ballistic basidiospores in the absence of dikaryon formation, diploid nuclei, and meiosis.

A rare few number of taxa have extended diploid lifecycles, but can be common species. Examples exist in the mushroom genera *Armillaria* and *Xerula*, both in the Physalacriaceae. Occasionally, basidiospores are not formed and parts of the "basidia" act as the dispersal agents, e.g., the peculiar mycoparasitic jelly fungus, *Tetragoniomyces* or the entire "basidium" acts as a "spore", e.g., in some false puffballs (*Scleroderma*). In the human pathogenic genus *Cryptococcus*, four nuclei following meiosis remain in the basidium, but continually divide mitotically, each nucleus migrating into synchronously forming nonballistic basidiospores that are then pushed upwards by another set forming below them, resulting in four parallel chains of dry "basidiospores". Other variations occur: some as standard lifecycles (that themselves have variations within variations) within specific orders.

Rusts

Rusts at their greatest complexity, produce five different types of spores on two different host plants in two unrelated host families. Such rusts are heteroecious (requiring two hosts) and macrocyclic (producing all five spores types). Wheat stem rust is an example. By convention, the stages and spore states are numbered by Roman numerals. Typically, basidiospores infect host one, also known as the alternate or sexual host, and the mycelium forms pycnidia, which are miniature, flask-shaped, hollow, submicroscopic bodies embedded in the host tissue (such as a leaf).

This stage, numbered "0", produces single-celled spores that ooze out in a sweet liquid and that act as nonmotile spermatia, and also protruding receptive hyphae. Insects and probably other vectors such as rain carry the spermatia from spermagonium to spermagonium, cross inoculating the mating types. Neither thallus is male or female. Once crossed, the dikaryons are established and a second spore stage is formed, numbered "I" and called aecia, which form dikaryotic aeciospores in dry chains in inverted cup-shaped bodies embedded in host tissue. These aeciospores then infect the second host, known as the primary or asexual host (in macrocyclic rusts). On the primary host a repeating spore stage is formed, numbered "II", the urediospores in dry pustules called uredinia. Urediospores are dikaryotic and can infect the same host that produced them. They repeatedly infect this host over the growing season. At the end of the season, a fourth spore type, the teliospore, is formed.

It is thicker-walled and serves to overwinter or to survive other harsh conditions. It does not continue the infection process, rather it remains dormant for a period and then germinates to form basidia (stage "IV"), sometimes called a promycelium. In the Pucciniales, the basidia are cylindrical and become 3-septate after meiosis, with each of the 4 cells bearing one basidiospore each. The basidiospores disperse and start the infection process on host 1 again. Autoecious rusts complete their life-cycles on one host instead of two, and microcyclic rusts cut out one or more stages.

Smuts

The characteristic part of the life-cycle of smuts is the thick-walled, often darkly pigmented, ornate, teliospore that serves to survive harsh conditions such as overwintering and also serves to help disperse the fungus as dry diaspores. The teliospores are initially dikaryotic but become diploid via karyogamy. Meiosis takes place at the time of germination. A promycelium is formed that consists of a short hypha (equated to a basidium). In some smuts such as *Mycosarcoma maydis* the nuclei migrate into the promycelium that becomes septate (i.e., divided into cellular compartments separated by cell walls called *septa*), and haploid yeast-like conidia/basidiospores sometimes called sporidia, bud off laterally from each cell.

In various smuts, the yeast phase may proliferate, or they may fuse, or they may infect plant tissue and become hyphal. In other smuts, such as *Tilletia caries*, the elongated haploid basidiospores form apically, often in compatible pairs that fuse centrally resulting in H-shaped diaspores which are by then dikaryotic. Dikaryotic conidia may then form. Eventually the host is infected by infectious hyphae. Teliospores form in host tissue. Many variations on these general themes occur.

Smuts with both a yeast phase and an infectious hyphal state are examples of dimorphic Basidiomycota. In plant parasitic taxa, the saprotrophic phase is normally the yeast while the infectious stage is hyphal. However, there are examples of animal and human parasites where the species are dimorphic but it is the yeast-like state that is infectious. The genus *Filobasidiella* forms basidia on hyphae but the main infectious stage is more commonly known by the anamorphic yeast name *Cryptococcus*, e.g. *Cryptococcus neoformans* and *Cryptococcus gattii*.

The dimorphic Basidiomycota with yeast stages and the pleiomorphic rusts are examples of fungi with anamorphs, which are the asexual stages. Some Basidiomycota are only known as anamorphs. Many are called basidiomycetous yeasts, which differentiates them from ascomycetous yeasts in the Ascomycota. Aside from yeast anamorphs and uredinia, aecia, and pycnidia, some Basidiomycota form other distinctive anamorphs as parts of their life cycles. Examples are *Collybia tuberosa* with its apple-seed-shaped and coloured sclerotium, *Dendrocollybia racemosa* with its sclerotium and its *Tilachlidiopsis racemosa* conidia, *Armillaria* with their rhizomorphs, *Hohenbuehelia* with their *Nematoctonus* nematode infectious, state and the coffee leaf parasite, *Mycena citricolor*, and its *Decapitatus flavidus* propagules called gemmae.

4.2.4 Chytridiomycota

Chytridiomycota are a division of zoosporic organisms in the kingdom Fungi, informally known as chytrids. Chytrids are one of the earliest diverging fungal lineages, and their membership in kingdom Fungi is demonstrated with chitin cell walls, a posterior whiplash flagellum, absorptive nutrition, use of glycogen as an energy storage compound, and synthesis of lysine by the α-amino adipic acid (AAA) pathway.

Chytrids are saprobic, degrading refractory materials such as chitin and keratin, and sometimes act as parasites. There has been a significant increase in the research of chytrids since the discovery of *Batrachochytrium dendrobatidis*, the causal agent of chytridiomycosis.

Classification

Species of Chytridiomycota have traditionally been delineated and classified based on development, morphology, substrate, and method of zoospore discharge. However, single spore isolates (or isogenic lines) display a great amount of variation in many of these features; thus, these features cannot be used to reliably classify or identify a species. Currently, taxonomy in Chytridiomycota is based on molecular data, zoospore ultrastructure and some aspects of thallus morphology and development.

In an older and more restricted sense (not used here), the term "chytrids" referred just to those fungi in the class Chytridiomycetes. Here, the term "chytrid" refers to all members of Chytridiomycota.

The chytrids have also been included among the Protoctista, but are now regularly classed as fungi. In older classifications, chytrids, except the recently established order Spizellomycetales, were placed in the class Phycomycetes under the subphylum Myxomycophyta of the kingdom Fungi. Previously, they were placed in the Mastigomycotina as the class Chytridiomycetes. The other classes of the Mastigomycotina, the Hyphochytriomycetes and oomycetes, were removed from the fungi to be classified as heterokont pseudofungi.

The class Chytridiomycetes has over 750 chytrid species distributed among ten orders. Additional classes include the Monoblepharidomycetes, with two orders, and the Hyaloraphidiomycetes with a single order. Molecular phylogenetics, and other techniques such as ultrastructure analysis, has greatly increased the understanding of chytrid phylogeny, and led to the formation of several new zoosporic fungal phyla:

- The order Blastocladiales, originally within the Chytridiomycota, are now classified as a separate phylum, the Blastocladiomycota.
- The Neocallimastigales, originally an order of anaerobic fungi of the class Chytridiomycetes, found in the digestive tracts of herbivores, was later raised to a separate phylum, the Neocallimastigomycota.
- The Olpidiaceae, including the type genus *Olpidium*, formerly classified in the order Chytridiales, were raised to a separate phylum, the Olpidiomycota.

Life Cycle

Chytridiomycota are unusual among the Fungi in that they reproduce with zoospores. For most members of Chytridiomycota, sexual reproduction is not known. Asexual reproduction occurs through the release of zoospores (presumably) derived through mitosis.

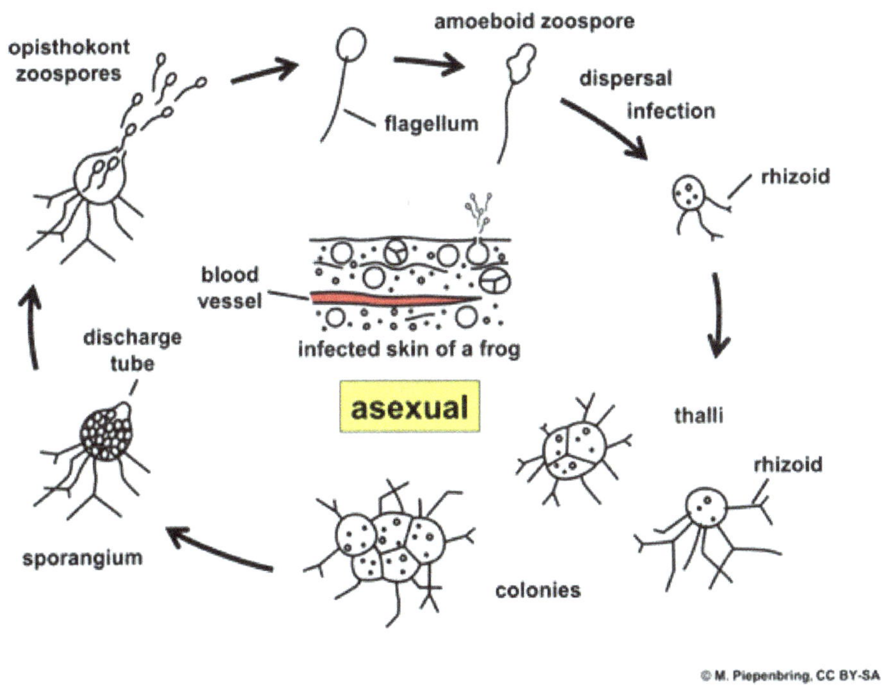

Fig 4: **Life Cycle of** *Batrachochytrium Dendrobatidis*

Where it has been described, sexual reproduction of chytrids occurs via a variety of methods. It is generally accepted that the resulting zygote forms a resting spore, which functions as a means of surviving adverse conditions. In some members, sexual reproduction is achieved through the fusion of isogametes (gametes of the same size and shape). This group includes the notable plant pathogens

Synchytrium. Some algal parasites practice oogamy: A motile male gamete attaches itself to a nonmotile structure containing the female gamete. In another group, two thalli produce tubes that fuse and allow the gametes to meet and fuse. In the last group, rhizoids of compatible strains meet and fuse. Both nuclei migrate out of the zoosporangium and into the conjoined rhizoids where they fuse. The resulting zygote germinates into a resting spore.

Sexual reproduction is common and well known among members of the Monoblepharidomycetes. Typically, these chytrids practice a version of oogamy: The male is motile and the female is stationary. This is the first occurrence of oogamy in kingdom Fungi. Briefly, the monoblephs form oogonia, which give rise to eggs, and antheridia, which give rise to male gametes. Once fertilized, the zygote either becomes an encysted or motile oospore, which ultimately becomes a resting spore that will later germinate and give rise to new zoosporangia.

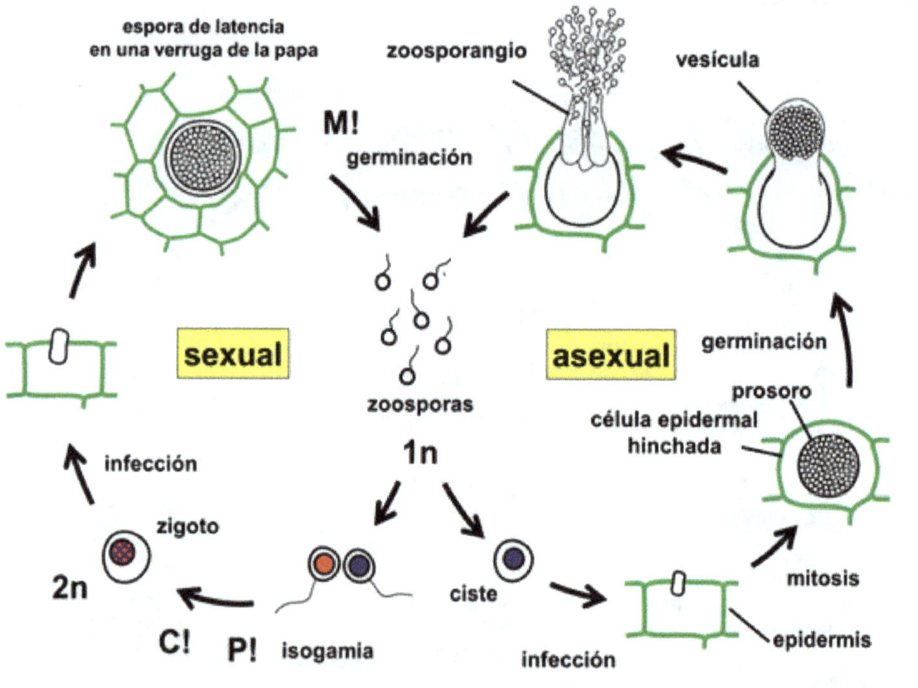

Fig 5: Life Cycle of *Synchytrium Endobioticum* in Potato

Upon release from the germinated resting spore, zoospores seek out a suitable substrate for growth using chemotaxis or phototaxis. Some species encyst and germinate directly upon the substrate; others encyst and germinate a short distance away. Once germinated, enzymes released from the zoospore begin to break down the substrate and utilize it produce a new thallus. Thalli are coenocytic and usually form no true mycelium (having rhizoids instead).

Fig 6: Synchytrium Endobioticum on Potatoes

Chytrids have several different growth patterns. Some are holocarpic, which means they only produce a zoosporangium and zoospores. Others are eucarpic, meaning they produce other structures, such as rhizoids, in addition to the zoosporangium and zoospores. Some chytrids are monocentric, meaning a single zoospore gives rise to a single zoosporangium. Others are polycentric, meaning one zoospore gives rise to many zoosporangium connected by a rhizomycelium. Rhizoids do not have nuclei while a rhizomycelium can.

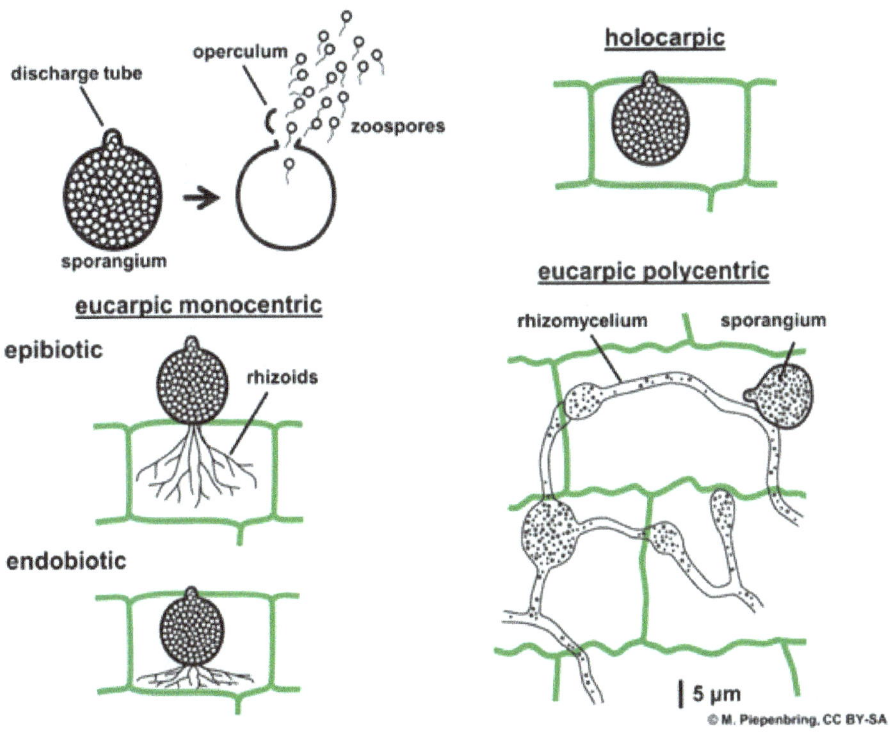

Fig 7: Types of Chytrid Thalli

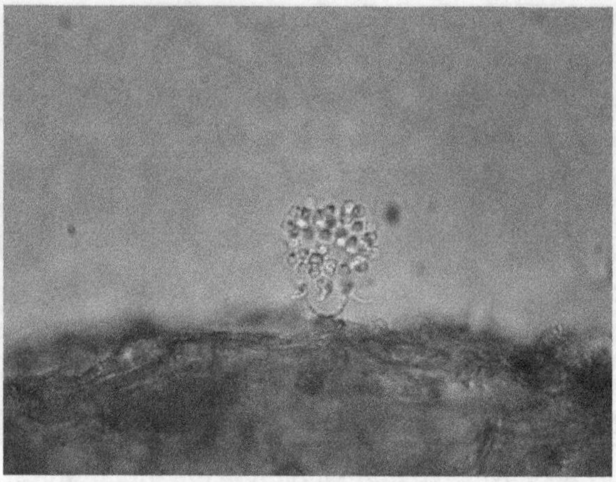

Fig 8: Zoospore Discharge from ***Phylctochytrium*** **sp.**

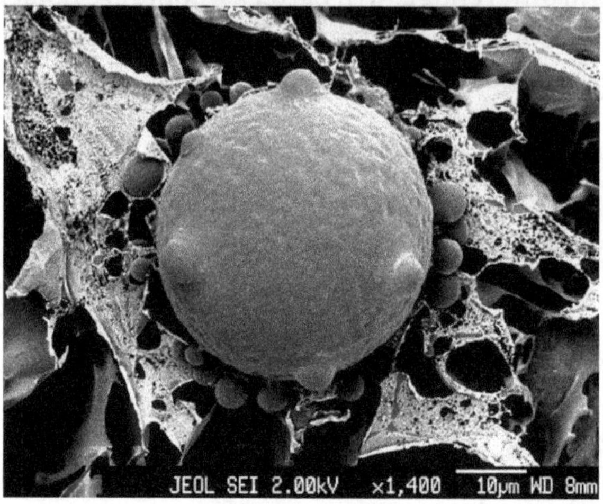

Fig 9: Sporangium and Zoospores of the Chytrid Fungus *B. Dendrobatidis*, **under a Scanning Electron Microscope**

Growth continues until a new batch of zoospores are ready for release. Chytrids have a diverse set of release mechanisms that can be grouped into the broad categories of operculate or inoperculate. Operculate discharge involves the complete or incomplete detachment of a lid-like structure, called an operculum, allowing the zoospores out of the sporangium. Inoperculate chytrids release their zoospores through pores, slits, or papillae.

Habitats

Chytrids are aquatic fungi, though those that thrive in the capillary network around soil particles are typically considered terrestrial. The zoospore is primarily a means of thoroughly exploring a small volume of water for a suitable substrate rather than a means of long-range dispersal. Chytrids have been isolated from a variety of aquatic habitats, including peats, bogs, rivers, ponds, springs, and ditches, and terrestrial habitats, such as acidic soils, alkaline soils, temperate forest soils, rainforest soils, Arctic and Antarctic soils. This has led to the belief that many chytrid species are ubiquitous and cosmopolitan. However, recent taxonomic work has demonstrated that this ubiquitous and cosmopolitan morphospecies hide cryptic diversity at the genetic and ultrastructural levels. It was first thought aquatic chytrids (and other zoosporic fungi) were primarily active in fall, winter, and spring. However, recent molecular inventories of lakes during the summer indicate that chytrids are an active, diverse part of the eukaryotic microbial community.

One of the least expected terrestrial environments the chytrid thrive in are periglacial soils. The population of the Chytridiomycota species are able to be supported even though there is a lack of plant life in these frozen regions due to the large amounts of water in periglacial soil and pollen blowing up from below the timberline.

Ecological Functions

Batrachochytrium Dendrobatidis

The chytrid *Batrachochytrium dendrobatidis* is responsible for chytridiomycosis, a disease of amphibians. Discovered in 1998 in Australia and Panama this disease is known to kill amphibians in large numbers, and has been suggested as a principal cause for the worldwide amphibian decline. Outbreaks of the fungus were found responsible for killing much of the Kihansi Spray Toad population in its native habitat of Tanzania, as well as the extinction of the golden toad in 1989. Chytridiomycosis has also been implicated in the presumed extinction of the Southern Gastric Brooding Frog, last seen in the wild in 1981, and the Northern Gastric Brooding Frog, last recorded in the wild in March 1985. The process leading to frog mortality is thought to be the loss of essential ions through pores made in the epidermal cells by the chytrid during its replication.

Recent research has revealed that elevating salt levels slightly may be able to cure chytridiomycosis in some Australian frog species, although further experimentation is needed.

Other Parasites

Chytrids mainly infect algae and other eukaryotic and prokaryotic microbes. The infection can be so severe as to control primary production within the lake. It has been suggested that parasitic chytrids have a large effect on lake and pond food webs. Chytrids may also infect plant species; in particular, *Synchytrium endobioticum* is an important potato pathogen.

Saprobes

Arguably, the most important ecological function chytrids perform is decomposition.

These ubiquitous and cosmopolitan organisms are responsible for decomposition of refractory materials, such as pollen, cellulose, chitin, and keratin. There are also chytrids that live and grow on pollen by attaching threadlike structures, called rhizoids, onto the pollen grains. This mostly occurs during asexual reproduction because the zoospores that become attached to the pollen continuously reproduce and form new chytrids that will attach to other pollen grains for nutrients. This colonization of pollen happens during the spring time when bodies of water accumulate pollen falling from trees and plants.

Fossil Record

The earliest fossils of chytrids are from the Scottish Rhynie chert, a Devonian-age lagerstätte with anatomical preservation of plants and fungi. Among the microfossils are chytrids preserved as parasites on rhyniophytes. These fossils closely resemble the modern genus *Allomyces*. Holocarpic chytrid remains were found in cherts from Combres in central France that date back to the late Visean. These remains were found along with eucarpic remains and are ambiguous in nature although they are thought to be of chytrids. Other chytrid-like fossils were found in cherts from the upper Pennsylvanian in the Saint-Etienne Basin in France, dating between 300~350 ma.

4.2.5 Glomeromycota

Glomeromycota (often referred to as glomeromycetes, as they include only one class, Glomeromycetes) are one of eight currently recognized divisions within the kingdom Fungi, with approximately 230 described species. Members of the Glomeromycota form arbuscular mycorrhizas (AMs) with the thalli of bryophytes and the roots of vascular land plants. Not all species have been shown to form AMs, and one, *Geosiphon pyriformis*, is known not to do so. Instead, it forms an endocytobiotic association with *Nostoc* cyanobacteria. The majority of evidence shows that the Glomeromycota are dependent on land plants (*Nostoc* in the case of *Geosiphon*) for carbon and energy, but there is recent circumstantial evidence that some species may be able to lead an independent existence. The arbuscular mycorrhizal species are terrestrial and widely distributed in soils worldwide where they form symbioses with the roots of the majority of plant species (>80%). They can also be found in wetlands, including salt-marshes, and associated with epiphytic plants. According to multigene phylogenetic analyses, this taxon is located as a member of the phylum Mucoromycota. Currently, the phylum name Glomeromycota is invalid, and the subphylum Glomeromycotina should be used to describe this taxon.

Reproduction

The Glomeromycota have generally coenocytic (occasionally sparsely septate) mycelia and reproduce asexually through blastic development of the hyphal tip to produce spores (Glomerospores,blastospore) with diameters of 80–500 μm. In some, complex spores form within a terminal saccule. Recently it was shown that *Glomus* species contain 51 genes encoding all the tools necessary for meiosis. Based on these and related findings, it was suggested that *Glomus* species may have a cryptic sexual cycle.

Colonization

New colonization of AM fungi largely depends on the amount of inoculum present in the soil. Although pre-existing hyphae and infected root fragments have been shown to colonize the roots of a host successfully, germinating spores are considered to be the key players in new host establishment. Spores are commonly dispersed by fungal and plant burrowing herbivore partners, but some air dispersal capabilities are also known. Studies have shown that spore germination is specific to particular environmental conditions such as right amount of nutrients, temperature or host availability. It has also been observed that the rate of root system colonization is directly correlated to spore density in the soil. In addition, new data also suggests that AM fungi host plants also secrete chemical factors that attract and enhance the growth of developing spore hyphae towards the root system.

The necessary components for the colonization of Glomeromycota include the host's fine root system, proper development of intracellular arbuscular structures, and a well-established external fungal mycelium. Colonization is accomplished by the interactions between germinating spore hyphae and the root hairs of the host or by the development of appressoria between epidermal root cells. The process is regulated by specialized chemical signaling and changes in gene expression of both the host and AM fungi. Intracellular hyphae extend up to the cortical cells of the root and penetrate the cell walls but not the inner cellular membrane creating an internal invagination. The penetrating hyphae develop a highly branched structure called an arbuscule, which has low functional periods before degradation and absorption by the host's root cells. A fully developed arbuscular mycorrhizal structure facilitates the two-way movement of nutrients between the host and mutualistic fungal partner. The symbiotic association allows the host plant to respond better to environmental stresses, and the non-photosynthetic fungi to obtain carbohydrates produced by photosynthesis.

Phylogeny

Initial studies of the Glomeromycota were based on the morphology of soil-borne sporocarps (spore clusters) found in or near colonized plant roots. Distinguishing features such as wall morphologies, size, shape, color, hyphal attachment and reaction to staining compounds allowed a phylogeny to be constructed. Superficial similarities led to the initial placement of genus *Glomus* in the unrelated family Endogonaceae. Following broader reviews that cleared up the sporocarp confusion, the Glomeromycota were first proposed in the genera *Acaulospora* and *Gigaspora* before being accorded their own order with the three families Glomaceae (now Glomeraceae), Acaulosporaceae and Gigasporaceae.

With the advent of molecular techniques this classification has undergone major revision. An analysis of small subunit (SSU) rRNA sequences indicated that they share a common ancestor with the Dikarya. Nowadays it is accepted that Glomeromycota consists of 4 orders.

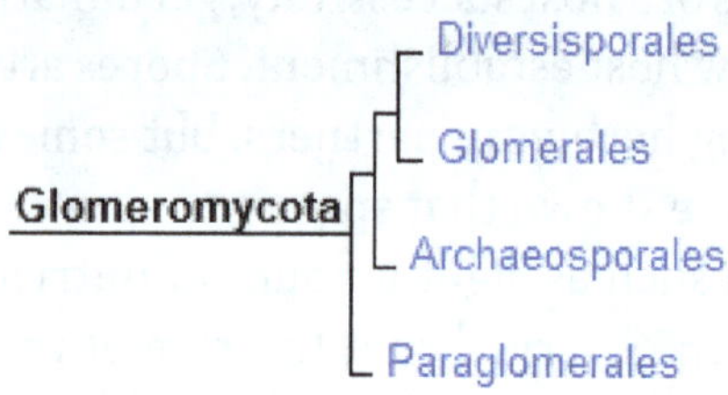

Several species which produce glomoid spores (i.e., spores similar to *Glomus*) in fact belong to other deeply divergent lineages and were placed in the orders, Paraglomerales and Archaeosporales.

This new classification includes the Geosiphonaceae, which presently contains one fungus (*Geosiphon pyriformis*) that forms endosymbiotic associations with the cyanobacterium *Nostoc punctiforme* and produces spores typical to this division, in the Archaeosporales. Work in this field is incomplete, and members of *Glomus* may be better suited to different genera or families.

Molecular biology

The biochemical and genetic characterization of the Glomeromycota has been hindered by their biotrophic nature, which impedes laboratory culturing. This obstacle was eventually surpassed with the use of root cultures and, most recently, a method which applies sequencing of single nucleus from spores has also been developed to circumvent this challenge. The first mycorrhizal gene to be sequenced was the small-subunit ribosomal RNA (SSU rRNA).

This gene is highly conserved and commonly used in phylogenetic studies so was isolated from spores of each taxonomic group before amplification through the polymerase chain reaction (PCR). A metatranscriptomic survey of the Sevilleta Arid Lands found that 5.4% of the fungal rRNA reads mapped to Glomeromycota.

This result was inconsistent with previous PCR-based studies of community structure in the region, suggesting that previous PCR-based studies may have underestimated Glomeromycota abundance due to amplification biases.

4.3 THE ECOLOGICAL ROLE OF FUNGI

Fungi play indispensable roles in ecosystems, serving as decomposers, symbionts, and pathogens.

- **Decomposition and Nutrient Cycling**: Fungi are among the primary decomposers in terrestrial ecosystems. By breaking down dead organic matter, such as fallen leaves, wood, and animal remains, fungi release essential nutrients back into the soil. This decomposition process supports plant growth and maintains the fertility of ecosystems.
- **Symbiotic Relationships**: Fungi engage in mutualistic relationships, the most famous of which is mycorrhizae. Mycorrhizal fungi colonize plant roots, helping them absorb water and nutrients, especially phosphorus. In return, plants supply fungi with carbohydrates produced through photosynthesis. Another example is lichens, a mutualistic partnership between fungi and photosynthetic algae or cyanobacteria. Lichens colonize bare rocks and help in soil formation.
- **Pathogens and Disease**: Fungi can also act as pathogens, infecting plants, animals, and even humans. Plant diseases like rusts, smuts, and blights caused by fungi can devastate crops. In humans, fungi can cause infections such as athlete's foot, ringworm, and more serious systemic infections in immunocompromised individuals.

4.4 ECONOMIC IMPORTANCE OF FUNGI

Fungi are highly valuable in various industries, including food, medicine, and agriculture. Their applications span from food production to pharmaceuticals and biotechnology.

- **Food Production**: Fungi are integral to the production of many foods and beverages. Yeasts, specifically *Saccharomyces cerevisiae*, are used in bread-making and fermentation processes for beer and wine production. Edible fungi, such as mushrooms and truffles, are delicacies with significant commercial value.
- **Pharmaceuticals**: Fungi have revolutionized medicine with the discovery of antibiotics. Penicillin, derived from the mold *Penicillium notatum*, was the first antibiotic and is one of the most important drugs in medical history. Other fungi produce immunosuppressants (e.g., cyclosporine) and cholesterol-lowering drugs (e.g., lovastatin). Fungi also serve as sources of anticancer and antiviral compounds, underscoring their pharmaceutical importance.

- **Biotechnology and Industrial Applications:** In biotechnology, fungi are used for the production of enzymes, organic acids, and biofuels. Enzymes from fungi facilitate processes in industries such as paper, textiles, and detergents. Fungi are also explored for biofuel production, as they can decompose lignocellulosic biomass and release sugars for fermentation into biofuels.
- **Agriculture:** Fungi are used as biocontrol agents to manage pests and diseases. Certain fungi, like *Trichoderma* species, protect plants from harmful fungi, reducing the need for chemical pesticides. Mycorrhizal fungi also support sustainable agriculture by improving nutrient uptake in crops, reducing the reliance on fertilizers.

4.5 CHALLENGES POSED BY FUNGI

While fungi provide significant benefits, they also pose challenges to human health, agriculture, and the environment.

- **Human Diseases:** Some fungi are pathogens that cause infections in humans, especially in those with compromised immune systems. Fungal infections can range from superficial conditions like ringworm to life-threatening systemic infections like aspergillosis and cryptococcosis. Managing fungal infections is complicated by the similarity between fungal and human cells, which makes developing effective antifungal drugs challenging.
- **Crop Losses and Food Security:** Fungal pathogens are a major threat to agriculture. Diseases caused by fungi, such as rusts, blights, and wilts, can severely reduce crop yields and affect food security. For example, wheat rust is a fungal disease that has led to large-scale crop losses worldwide. The spread of fungal pathogens is exacerbated by global trade and climate change.
- **Antifungal Resistance:** Like bacteria, fungi can develop resistance to antifungal treatments. Antifungal resistance complicates the treatment of fungal infections and poses a significant threat to public health. Overuse of antifungal agents in agriculture and medicine contributes to the rise of resistant fungal strains.

4.6 RECENT ADVANCES IN FUNGAL RESEARCH

Advancements in fungal research have expanded our understanding of these organisms and their applications. Genomics, molecular biology, and environmental science have enabled scientists to explore fungal diversity, ecology, and biotechnology.

- **Fungal Genomics:** Genome sequencing has provided insights into the diversity and metabolic pathways of fungi, uncovering genes involved in the production of bioactive compounds, enzymes, and other valuable substances. Genomic research facilitates the discovery of new drugs and industrial products.
- **Synthetic Biology:** Synthetic biology approaches are being used to engineer fungi for specific purposes, such as biofuel production or the synthesis of high-value chemicals. By manipulating fungal genes, researchers can optimize the production of compounds of interest and reduce reliance on traditional chemical processes.
- **Conservation and Biodiversity Studies:** As fungi are key to ecosystem health, research on fungal biodiversity and conservation is increasing. Environmental DNA (eDNA) studies are revealing previously unknown fungal species and improving our understanding of fungal roles in ecosystems.

4.7 CONCLUSION

Fungi are fascinating and versatile organisms that play essential roles in nature, industry, and human health. From their contributions to nutrient cycling and agriculture to their impact on medicine and biotechnology, fungi are indispensable to ecosystems and economies. Despite their benefits, fungi also present challenges in the form of diseases, agricultural threats, and antifungal resistance. Continued research is crucial to harness their potential while managing the risks they pose. As our understanding of fungi deepens, these organisms are likely to become even more central to sustainable development, innovative medicine, and environmental health.

CHAPTER 5
VIRUSES AND OTHER ACELLULAR INFECTIOUS AGENTS: CHARACTERISTICS AND CONTROL

Viruses and other acellular infectious agents, such as viroids, prions, and virus-like particles, are unique pathogens that stand apart from traditional cellular organisms.

Lacking cellular structures, these infectious agents depend entirely on a host to replicate, causing various diseases in plants, animals, and humans.

5.1 CHARACTERISTICS OF VIRUSES

Viruses are the most studied acellular infectious agents due to their significant impact on human health, agriculture, and animal populations. Unlike bacteria, fungi, or protozoa, viruses are composed mainly of genetic material encapsulated within a protein coat, sometimes surrounded by an additional lipid envelope.

- **Structure of Viruses**: Viruses consist of nucleic acids (either DNA or RNA) encased within a protein shell called a capsid. The capsid protects the viral genome and enables it to attach to and enter host cells. Some viruses, especially those infecting animals, have an outer lipid envelope derived from the host cell membrane. This envelope often contains viral proteins critical for recognizing and binding to host cells.
- **Genetic Diversity**: Viruses exhibit a wide range of genetic variations. Some viruses have single-stranded RNA (such as the influenza virus), while others have double-stranded DNA (like the herpesvirus). Their genomes can be linear or circular and are usually quite compact, encoding only the proteins essential for infection and replication within a host.
- **Replication and Host Dependence**: Viruses cannot reproduce independently. To replicate, they must infect a host cell, using its machinery to produce viral proteins and new viral particles. After entering the host cell, viral genetic material is expressed and replicated, often resulting in the destruction of the host cell as new viruses are released.

5.2 VIROIDS AND PRIONS: OTHER ACELLULAR INFECTIOUS AGENTS

While viruses are the most well-known acellular pathogens, viroids and prions are also significant agents of infection, particularly in plants and animals.

5.2.1 Viroids

Viroids are small single-stranded, circular RNAs that are infectious pathogens. Unlike viruses, they have no protein coating. All known viroids are inhabitants of angiosperms (flowering plants), and most cause diseases, whose respective economic importance to humans varies widely. A recent metatranscriptomics study suggests that the host diversity of viroids and other viroid-like elements is broader than previously thought and that it would not be limited to plants, encompassing even the prokaryotes. The first discoveries of viroids in the 1970s triggered the historically third major extension of the biosphere—to include smaller lifelike entities—after the discoveries in 1675 by Antonie van Leeuwenhoek (of the "subvisible" microorganisms) and in 1892–1898 by Dmitri Iosifovich Ivanovsky and Martinus Beijerinck (of the "submicroscopic" viruses). The unique properties of viroids have been recognized by the International Committee on Taxonomy of Viruses, in creating a new order of subviral agents.

The first recognized viroid, the pathogenic agent of the potato spindle tuber disease, was discovered, initially molecularly characterized, and named by Theodor Otto Diener, plant pathologist at the U.S Department of Agriculture's Research Center in Beltsville, Maryland, in 1971. This viroid is now called potato spindle tuber viroid, abbreviated PSTVd. The *Citrus exocortis viroid* (CEVd) was discovered soon thereafter, and together understanding of PSTVd and CEVd shaped the concept of the viroid.

Although viroids are composed of nucleic acid, they do not code for any protein. The viroid's replication mechanism uses RNA polymerase II, a host cell enzyme normally associated with synthesis of messenger RNA from DNA, which instead catalyzes "rolling circle" synthesis of new RNA using the viroid's RNA as a template. Viroids are often ribozymes, having catalytic properties that allow self-cleavage and ligation of unit-size genomes from larger replication intermediates.

Diener initially hypothesized in 1989 that viroids may represent "living relics" from the widely assumed, ancient, and non-cellular RNA world, and others have followed this conjecture. Following the discovery of retrozymes, it has been proposed that viroids and other viroid-like elements may derive from this newly found class of retrotransposon. The human pathogen hepatitis D virus is a subviral agent similar in structure to a viroid, as it is a hybrid particle enclosed by surface proteins from the hepatitis B virus.

History

In the 1920s, symptoms of a previously unknown potato disease were noticed in New York and New Jersey fields. Because tubers on affected plants become elongated and misshapen, they named it the potato spindle tuber disease.

The symptoms appeared on plants onto which pieces from affected plants had been budded—indicating that the disease was caused by a transmissible pathogenic agent. A fungus or bacterium could not be found consistently associated with symptom-bearing plants, however, and therefore, it was assumed the disease was caused by a virus. Despite numerous attempts over the years to isolate and purify the assumed virus, using increasingly sophisticated methods, these were unsuccessful when applied to extracts from potato spindle tuber disease-afflicted plants.

In 1971, Theodor O. Diener showed that the agent was not a virus, but a totally unexpected novel type of pathogen, 1/80th the size of typical viruses, for which he proposed the term "viroid". Parallel to agriculture-directed studies, more basic scientific research elucidated many of viroids' physical, chemical, and macromolecular properties.

Viroids were shown to consist of short stretches (a few hundred nucleotides) of single-stranded RNA and, unlike viruses, did not have a protein coat. Viroids are extremely small, from 246 to 467 nucleotides, smaller than other infectious plant pathogens; they thus consist of fewer than 10,000 atoms. In comparison, the genomes of the smallest known viruses capable of causing an infection by themselves are around 2,000 nucleotides long.

In 1976, Sanger, *et al.*, presented evidence that potato spindle tuber viroid is a "single-stranded, covalently closed, circular RNA molecule, existing as a highly base-paired rod-like structure"—believed to be the first such molecule described.

Circular RNA, unlike linear RNA, forms a covalently closed continuous loop, in which the 3' and 5' ends present in linear RNA molecules have been joined. Sanger et al. also provided evidence for the true circularity of viroids by finding that the RNA could not be phosphorylated at the 5' terminus. In other tests, they failed to find even one free 3' end, which ruled out the possibility of the molecule having two 3' ends. Viroids thus are true circular RNAs.

The single-strandedness and circularity of viroids was confirmed by electron microscopy, The complete nucleotide sequence of potato spindle tuber viroid was determined in 1978. PSTVd was the first pathogen of a eukaryotic organism for which the complete molecular structure has been established. Over thirty plant diseases have since been identified as viroid-, not virus-caused, as had been assumed.

Transmission and Replication

Viroids are only known to infect plants, and infectious viroids can be transmitted to new plant hosts by aphids, by cross contamination following mechanical damage to plants as a result of horticultural or agricultural practices, or from plant to plant by leaf contact. Upon infection, viroids replicate in the nucleus (*Pospiviroidae*) or chloroplasts (*Avsunviroidae*) of plant cells in three steps through an RNA-based mechanism.

They require RNA polymerase II, a host cell enzyme normally associated with synthesis of messenger RNA from DNA, which instead catalyzes "rolling circle" synthesis of new RNA using the viroid as template. Unlike plant viruses which produce movement proteins, viroids are entirely passive, relying entirely on the host. This is useful in the study of RNA kinetics in plants.

RNA Silencing

There has long been uncertainty over how viroids induce symptoms in plants without encoding any protein products within their sequences. Evidence suggests that RNA silencing is involved in the process. First, changes to the viroid genome can dramatically alter its virulence. This reflects the fact that any siRNAs produced would have less complementary base pairing with target messenger RNA. Secondly, siRNAs corresponding to sequences from viroid genomes have been isolated from infected plants.

Finally, transgenic expression of the noninfectious hpRNA of potato spindle tuber viroid develops all the corresponding viroid-like symptoms. This indicates that when viroids replicate via a double stranded intermediate RNA, they are targeted by a dicer enzyme and cleaved into siRNAs that are then loaded onto the RNA-induced silencing complex. The viroid siRNAs contain sequences capable of complementary base pairing with the plant's own messenger RNAs, and induction of degradation or inhibition of translation causes the classic viroid symptoms.

Viroid-Like Elements

"Viroid-like elements" refer to pieces of covalently closed circular (ccc) RNA molecules that do not share the viroid's lifecycle. The category encompasses satellite RNAs (including small plant satRNAs "virusoids", fungal "ambivirus", and the much larger HDV-like *Ribozyviria*) and "retroviroids". Most of them also carry some type of a ribozyme.

Viroid-Like Satellite RNAs

Viroid-like satellite RNAs are infectious circular RNA molecules that depend on a carrier virus to reproduce, being carried in their capsids. Like Avsunviroidae, however, they are capable of self-clevage.

Ambiviruses

"Ambiviruses" are mobile genetic elements that were recently (2020s) discovered in fungi. Their RNA genomes are circular, circa 5 kb in length. One of at least two open reading frames encodes a viral RNA-directed RNA polymerase, that firmly places "ambiviruses" into ribovirian kingdom *Orthornavirae*; a separate phylum *Ambiviricota* has been established since the 2023 ICTV Virus Taxonomy Release because of the unique features of encoding RNA-directed RNA polymerases but also having divergent ribozymes in various combinations in both sense and antisense orientation – the detection of circular forms in both sense orientations suggest that "ambiviruses" use rolling circle replication for propagation.

Retroviroids

"Retroviroids", more formally "retroviroid-like elements", are viroid-like circular RNA sequences that are also found with homologous copies in the DNA genome of the host. The only types found are closely related to the original "carnation small viroid-like RNA" (CarSV). These elements may act as a homologous substrate upon which recombination may occur and are linked to double-stranded break repair. These elements are dubbed retroviroids as the homologous DNA is generated by reverse transcriptase that is encoded by retroviruses. They are neither true viroids nor viroid-like satellite RNAs: there is no extracellular form of these elements; instead, they are spread only through pollen or egg-cells. They appear to co-occur with a pararetrovirus.

Obelisks

After applying metatranscriptomics–the computer-aided search for RNA sequences and their analysis – biologists reported in January 2024 the discovery of "obelisks", a new class of viroid-like elements, and "oblins", their related group of proteins, in the human microbiome. Given that the RNA sequences recovered do not have homologies in any other known life form, the researchers suggest that the obelisks are distinct from viruses, viroids and viroid-like entities, and thus form an entirely new class of organisms.

RNA World Hypothesis

Diener's 1989 hypothesis had proposed that the unique properties of viroids make them more plausible macromolecules than introns, or other RNAs considered in the past as possible "living relics" of a hypothetical, pre-cellular RNA world.

If so, viroids have assumed significance beyond plant virology for evolutionary theory, because their properties make them more plausible candidates than other RNAs to perform crucial steps in the evolution of life from inanimate matter (abiogenesis).

Diener's hypothesis was mostly forgotten until 2014, by Flores, *et al.*, in which the authors summarized Diener's evidence supporting his hypothesis as:

- Viroids' small size, imposed by error-prone replication.
- Their high guanine and cytosine content, which increases stability and replication fidelity.
- Their circular structure, which assures complete replication without genomic tags.
- Existence of structural periodicity, which permits modular assembly into enlarged genomes.
- Their lack of protein-coding ability, consistent with a ribosome-free habitat.
- Replication mediated in some by ribozymes—the fingerprint of the RNA world.

The presence, in extant cells, of RNAs with molecular properties predicted for RNAs of the RNA world constitutes another powerful argument supporting the RNA world hypothesis. However, the origins of viroids themselves from this RNA world has been cast into doubt by several factors, including the discovery of retrozymes (a family of retrotransposon likely representing their ancestors) and their complete absence from organisms outside of the plants (especially their complete absence from prokaryotes including bacteria and archaea). However, recent studies suggest that the diversity of viroids and others viroid-like elements is broader than previously thought and that it would not be limited to plants, encompassing even the prokaryotes. Matches between viroid cccRNAs and CRISPR spacers suggest that some of them might replicate in prokaryotes.

Control

The development of tests based on ELISA, PCR, and nucleic acid hybridization has allowed for rapid and inexpensive detection of known viroids in biosecurity inspections, phytosanitary inspections, and quarantine.

5.2.2 Prions

A prion is a misfolded protein that induces misfolding in normal variants of the same protein, leading to cellular death. Prions are responsible for prion diseases, known as transmissible spongiform encephalopathy (TSEs), which are fatal and transmissible neurodegenerative diseases affecting both humans and animals. These proteins can misfold sporadically, due to genetic mutations, or by exposure to an already misfolded protein, leading to an abnormal three-dimensional structure that can propagate misfolding in other proteins.

The term *prion* comes from "proteinaceous infectious particle". Unlike other infectious agents such as viruses, bacteria, and fungi, prions do not contain nucleic acids (DNA or RNA). Prions are mainly twisted isoforms of the major prion protein (PrP), a naturally occurring protein with an uncertain function. They are the hypothesized cause of various TSEs, including scrapie in sheep, chronic wasting disease (CWD) in deer, bovine spongiform encephalopathy (BSE) in cattle (mad cow disease), and Creutzfeldt–Jakob disease (CJD) in humans.

All known prion diseases in mammals affect the structure of the brain or other neural tissues. These diseases are progressive, have no known effective treatment, and are invariably fatal. Most prion diseases were thought to be caused by PrP until 2015 when a prion form of alpha-synuclein was linked to multiple system atrophy (MSA). Prions are also linked to other neurodegenerative diseases like Alzheimer's disease, Parkinson's disease, and amyotrophic lateral sclerosis (ALS), which are sometimes referred to as *prion-like diseases.*

Prions are a type of intrinsically disordered protein that continuously changes conformation unless bound to a specific partner, such as another protein. Once a prion binds to another in the same conformation, it stabilizes and can form a fibril, leading to abnormal protein aggregates called amyloids. These amyloids accumulate in infected tissue, causing damage and cell death. The structural stability of prions makes them resistant to denaturation by chemical or physical agents, complicating disposal and containment, and raising concerns about iatrogenic spread through medical instruments.

History

In the 18th and 19th centuries, exportation of sheep from Spain was observed to coincide with a disease called scrapie. This disease caused the affected animals to *"lie down, bite at their feet and legs, rub their backs against posts, fail to thrive, stop feeding and finally become lame"*. The disease was also observed to have the long incubation period that is a key characteristic of transmissible spongiform encephalopathies (TSEs). Although the cause of scrapie was not known back then, it is probably the first transmissible spongiform encephalopathy to be recorded.

In the 1950s, Carleton Gajdusek began research which eventually showed that kuru could be transmitted to chimpanzees by what was possibly a new infectious agent, work for which he eventually won the 1976 Nobel prize. During the 1960s, two London-based researchers, radiation biologist Tikvah Alper and biophysicist John Stanley Griffith, developed the hypothesis that the transmissible spongiform encephalopathies are caused by an infectious agent consisting solely of proteins. Earlier investigations by E.J. Field into scrapie and kuru had found evidence for the transfer of pathologically inert polysaccharides that only become infectious post-transfer, in the new host. Alper and Griffith wanted to account for the discovery that the mysterious infectious agent causing the diseases scrapie and Creutzfeldt-Jakob disease resisted ionizing radiation. Griffith proposed three ways in which a protein could be a pathogen.

In the first hypothesis, he suggested that if the protein is the product of a normally suppressed gene, and introducing the protein could induce the gene's expression, that is, wake the dormant gene up, then the result would be a process indistinguishable from replication, as the gene's expression would produce the protein, which would then wake the gene in other cells.

His second hypothesis forms the basis of the modern prion theory, and proposed that an abnormal form of a cellular protein can convert normal proteins of the same type into its abnormal form, thus leading to replication.

His third hypothesis proposed that the agent could be an antibody if the antibody was its own target antigen, as such an antibody would result in more and more antibody being produced against itself. However, Griffith acknowledged that this third hypothesis was unlikely to be true due to the lack of a detectable immune response.

Francis Crick recognized the potential significance of the Griffith protein-only hypothesis for scrapie propagation in the second edition of his "Central dogma of molecular biology" (1970): While asserting that the flow of sequence information from protein to protein, or from protein to RNA and DNA was "precluded", he noted that Griffith's hypothesis was a potential contradiction (although it was not so promoted by Griffith). The revised hypothesis was later formulated, in part, to accommodate reverse transcription (which both Howard Temin and David Baltimore discovered in 1970).

In 1982, Stanley B. Prusiner of the University of California, San Francisco, announced that his team had purified the hypothetical infectious protein, which did not appear to be present in healthy hosts, though they did not manage to isolate the protein until two years after Prusiner's announcement. The protein was named a prion, for "proteinacious infectious particle", derived from the words protein and infection. When the prion was discovered, Griffith's first hypothesis, that the protein was the product of a normally silent gene was favored by many. It was subsequently discovered, however, that the same protein exists in normal hosts but in different form.

Following the discovery of the same protein in different form in uninfected individuals, the specific protein that the prion was composed of was named the prion protein (PrP), and Griffith's second hypothesis that an abnormal form of a host protein can convert other proteins of the same type into its abnormal form, became the dominant theory. Prusiner was awarded the Nobel Prize in Physiology or Medicine in 1997 for his research into prions.

Prion Protein

Structure

Prions consist of a misfolded form of major prion protein (PrP), a protein that is a natural part of the bodies of humans and other animals. The PrP found in infectious prions has a different structure and is resistant to proteases, the enzymes in the body that can normally break down proteins. The normal form of the protein is called PrP^{C}, while the infectious form is called PrP^{Sc} – the *C* refers to 'cellular' PrP, while the *Sc* refers to 'scrapie', the prototypic prion disease, occurring in sheep. PrP can also be induced to fold into other more-or-less well-defined isoforms in vitro; although their relationships to the form(s) that are pathogenic in vivo is often unclear, high-resolution structural analyses have begun to reveal structural features that correlate with prion infectivity.

PrPC

PrP^{C} is a normal protein found on the membranes of cells, "including several blood components of which platelets constitute the largest reservoir in humans." It has 209 amino acids (in humans), one disulfide bond, a molecular mass of 35–36 kDa and a mainly alpha-helical structure. Several topological forms exist; one cell surface form anchored via glycolipid and two transmembrane forms. The normal protein is not sedimentable; meaning that it cannot be separated by centrifuging techniques. It has a complex function, which continues to be investigated. PrP^{C} binds copper(II) ions (those in a +2 oxidation state) with high affinity. The significance of this property is not clear, but it is presumed to relate to the protein's structure or function. PrP^{C} is readily digested by proteinase K and can be liberated from the cell surface by the enzyme phosphoinositide phospholipase C (PI-PLC), which cleaves the glycophosphatidylinositol (GPI) glycolipid anchor. PrP plays an important role in cell-cell adhesion and intracellular signaling *in vivo*, and may therefore be involved in cell-cell communication in the brain.

PrPSc

The infectious isoform of PrP, known as PrP^{Sc}, or simply the prion, is able to convert normal PrP^{C} proteins into the infectious isoform by changing their conformation, or shape; this, in turn, alters the way the proteins interconnect. PrP^{Sc} always causes prion disease. PrP^{Sc} has a higher proportion of β-sheet structure in place of the normal α-helix structure. Several highly infectious, brain-derived PrP^{Sc} structures have been discovered by cryo-electron microscopy. Another brain-derived fibril structure isolated from humans with Gerstmann-Straussler-Schienker syndrome has also been determined. All of the structures described in high resolution so far are amyloid fibers in which individual PrP molecules are stacked via intermolecular beta sheets. However, 2-D crystalline arrays have also been reported at lower resolution in *ex vivo* preparations of prions. In the prion amyloids, the glycolipid anchors and asparagine-linked glycans, when present, project outward from the lateral surfaces of the fiber cores. Often PrP^{Sc} is bound to cellular membranes, presumably via its array of glycolipid anchors, however, sometimes the fibers are dissociated from membranes and accumulate outside of cells in the form of plaques. The end of each fiber acts as a template onto which free protein molecules may attach, allowing the fiber to grow.

This growth process requires complete refolding of PrP^{C}. Different prion strains have distinct templates, or conformations, even when composed of PrP molecules of the same amino acid sequence, as occurs in a particular host genotype. Under most circumstances, only PrP molecules with an identical amino acid sequence to the infectious PrP^{Sc} are incorporated into the growing fiber. However, cross-species transmission also happens rarely.

PrPres

Protease-resistant PrP^{Sc}-like protein (PrP^{res}) is the name given to any isoform of PrP^{c} which is structurally altered and converted into a misfolded proteinase K-resistant form. To model conversion of PrP^{C} to PrP^{Sc} *in vitro*, Kocisko *et al.* showed that PrP^{Sc} could cause PrP^{C} to convert to PrP^{res} under cell-free conditions and Soto *et al.* demonstrated sustained amplification of PrP^{res} and prion infectivity by a procedure involving cyclic amplification of protein misfolding. The term "PrP^{res}" may refer either to protease-resistant forms of PrP^{Sc}, which is isolated from infectious tissue and associated with the transmissible spongiform encephalopathy agent, or to other protease-resistant forms of PrP that, for example, might be generated *in vitro*. Accordingly, unlike PrP^{Sc}, PrP^{res} may not necessarily be infectious.

Normal Function of PrP

The physiological function of the prion protein remains poorly understood. While data from in vitro experiments suggest many dissimilar roles, studies on PrP knockout mice have provided only limited information because these animals exhibit only minor abnormalities. In research done in mice, it was found that the cleavage of PrP in peripheral nerves causes the activation of myelin repair in Schwann cells and that the lack of PrP proteins caused demyelination in those cells.

- **PrP and Regulated Cell Death:** MAVS, RIP1, and RIP3 are prion-like proteins found in other parts of the body. They also polymerise into filamentous amyloid fibers which initiate regulated cell death in the case of a viral infection to prevent the spread of virions to other, surrounding cells.
- **PrP and Long-Term Memory:** A review of evidence in 2005 suggested

that PrP may have a normal function in maintenance of long-term memory. As well, a 2004 study found that mice lacking genes for normal cellular PrP protein show altered hippocampal long-term potentiation. A recent study that also suggests why this might be the case, found that neuronal protein CPEB has a similar genetic sequence to yeast prion proteins. The prion-like formation of CPEB is essential for maintaining long-term synaptic changes associated with long-term memory formation.

- **PrP and Stem Cell Renewal:** A 2006 article from the Whitehead Institute for Biomedical Research indicates that PrP expression on stem cells is necessary for an organism's self-renewal of bone marrow. The study showed that all long-term hematopoietic stem cells express PrP on their cell membrane and that hematopoietic tissues with PrP-null stem cells exhibit increased sensitivity to cell depletion.
- **PrP and Innate Immunity:** There is some evidence that PrP may play a role in innate immunity, as the expression of PRNP, the PrP gene, is upregulated in many viral infections and PrP has antiviral properties against many viruses, including HIV.

Replication

The first hypothesis that tried to explain how prions replicate in a protein-only manner was the heterodimer model. This model assumed that a single PrP^{Sc} molecule binds to a single PrP^{C} molecule and catalyzes its conversion into PrP^{Sc}. The two PrP^{Sc} molecules then come apart and can go on to convert more PrP^{C}. However, a model of prion replication must explain both how prions propagate, and why their spontaneous appearance is so rare. Manfred Eigen showed that the heterodimer model requires PrP^{Sc} to be an extraordinarily effective catalyst, increasing the rate of the conversion reaction by a factor of around 10^{15}. This problem does not arise if PrP^{Sc} exists only in aggregated forms such as amyloid, where cooperativity may act as a barrier to spontaneous conversion. What is more, despite considerable effort, infectious monomeric PrP^{Sc} has never been isolated.

An alternative model assumes that PrP^{Sc} exists only as fibrils, and that fibril ends bind PrP^{C} and convert it into PrP^{Sc}. If this were all, then the quantity of prions would increase linearly, forming ever longer fibrils. But exponential growth of both PrP^{Sc} and of the quantity of infectious particles is observed during prion disease. This can be explained by taking into account fibril breakage. A mathematical solution for the exponential growth rate resulting from the combination of fibril growth and fibril breakage has been found. The exponential growth rate depends largely on the square root of the PrP^{C} concentration. The incubation period is determined by the exponential growth rate, and in vivo data on prion diseases in transgenic mice match this prediction. The same square root dependence is also seen in vitro in experiments with a variety of different amyloid proteins.

The mechanism of prion replication has implications for designing drugs. Since the incubation period of prion diseases is so long, an effective drug does not need to eliminate all prions, but simply needs to slow down the rate of exponential growth. Models predict that the most effective way to achieve this, using a drug with the lowest possible dose, is to find a drug that binds to fibril ends and blocks them from growing any further.

Researchers at Dartmouth College discovered that endogenous host cofactor molecules such as the phospholipid molecule (e.g., phosphatidylethanolamine) and polyanions (e.g., single stranded RNA molecules) are necessary to form PrP^{Sc} molecules with high levels of specific infectivity *in vitro*, whereas protein-only PrP^{Sc} molecules appear to lack significant levels of biological infectivity.

Transmissible Spongiform Encephalopathies

Prions cause neurodegenerative disease by aggregating extracellularly within the central nervous system to form plaques known as amyloids, which disrupt the normal tissue structure.

This disruption is characterized by "holes" in the tissue with resultant spongy architecture due to the vacuole formation in the neurons. Other histological changes include astrogliosis and the absence of an inflammatory reaction. While the incubation period for prion diseases is relatively long (5 to 20 years), once symptoms appear the disease progresses rapidly, leading to brain damage and death. Neurodegenerative symptoms can include convulsions, dementia, ataxia (balance and coordination dysfunction), and behavioural or personality changes.

Table 1: Diseases Caused by Prions

Affected Animal(s)	Disease
Sheep, Goat	Scrapie
Cattle	Bovine spongiform encephalopathy
Camel	Camel spongiform encephalopathy (CSE)
Mink	Transmissible mink encephalopathy (TME)
White-tailed deer, elk, mule deer, moose	Chronic wasting disease (CWD)
Cat	Feline spongiform encephalopathy (FSE)
Nyala, Oryx, Greater Kudu	Exotic ungulate encephalopathy (EUE)
Ostrich	Spongiform encephalopathy (unknown if transmissible)

Human	Creutzfeldt–Jakob disease (CJD)
	Iatrogenic Creutzfeldt–Jakob disease (iCJD)
	Variant Creutzfeldt–Jakob disease (vCJD)
	Familial Creutzfeldt–Jakob disease (fCJD)
	Sporadic Creutzfeldt–Jakob disease (sCJD)
	Gerstmann–Sträussler–Scheinker syndrome (GSS)
	Fatal insomnia (FFI)
	Kuru
	Familial spongiform encephalopathy
	Variably protease-sensitive prionopathy (VPSPr)

Many different mammalian species can be affected by prion diseases, as the prion protein (PrP) is very similar in all mammals. Due to small differences in PrP between different species it is unusual for a prion disease to transmit from one species to another. The human prion disease variant Creutzfeldt–Jakob disease, however, is thought to be caused by a prion that typically infects cattle, causing bovine spongiform encephalopathy and is transmitted through infected meat. All known prion diseases are untreatable and fatal. Until 2015 all known mammalian prion diseases were considered to be caused by the prion protein, PrP; in 2015 multiple system atrophy was found to be transmissible and was hypothesized to be caused by a new prion, the misfolded form of a protein called alpha-synuclein. The endogenous, properly folded form of the prion protein is denoted PrP^{C} (for *Common* or *Cellular*), whereas the disease-linked, misfolded form is denoted PrP^{Sc} (for *Scrapie*), after one of the diseases first linked to prions and neurodegeneration. The precise structure of the prion is not known, though they can be formed spontaneously by combining PrP^{C}, homopolymeric polyadenylic acid, and lipids in a protein misfolding cyclic amplification (PMCA) reaction even in the absence of pre-existing infectious prions. This result is further evidence that prion replication does not require genetic information.

Transmission

It has been recognized that prion diseases can arise in three different ways: acquired, familial, or sporadic. It is often assumed that the diseased form directly interacts with the normal form to make it rearrange its structure. One idea, the "Protein X" hypothesis, is that an as-yet unidentified cellular protein (Protein X) enables the conversion of PrP^{C} to PrP^{Sc} by bringing a molecule of each of the two together into a complex.

The primary method of infection in animals is through ingestion. It is thought that prions may be deposited in the environment through the remains of dead animals and via urine, saliva, and other body fluids. They may then linger in the soil by binding to clay and other minerals.

A University of California research team has provided evidence for the theory that infection can occur from prions in manure. And, since manure is present in many areas surrounding water reservoirs, as well as used on many crop fields, it raises the possibility of widespread transmission. Although it was initially reported in January 2011 that researchers had discovered prions spreading through airborne transmission on aerosol particles in an animal testing experiment focusing on scrapie infection in laboratory mice, this report was retracted in 2024. Preliminary evidence supporting the notion that prions can be transmitted through use of urine-derived human menopausal gonadotropin, administered for the treatment of infertility.

- **Prions in Plants:** In 2015, researchers at The University of Texas Health Science Center at Houston found that plants can be a vector for prions. When researchers fed hamsters grass that grew on ground where a deer that died with chronic wasting disease (CWD) was buried, the hamsters became ill with CWD, suggesting that prions can bind to plants, which then take them up into the leaf and stem structure, where they can be eaten by herbivores, thus completing the cycle. It is thus possible that there is a progressively accumulating number of prions in the environment.

Sterilization

Infectious particles possessing nucleic acid are dependent upon it to direct their continued replication. Prions, however, are infectious by their effect on normal versions of the protein. Sterilizing prions, therefore, requires the denaturation of the protein to a state in which the molecule is no longer able to induce the abnormal folding of normal proteins. In general, prions are quite resistant to proteases, heat, ionizing radiation, and formaldehyde treatments, although their infectivity can be reduced by such treatments. Effective prion decontamination relies upon protein hydrolysis or reduction or destruction of protein tertiary structure. Examples include sodium hypochlorite, sodium hydroxide, and strongly acidic detergents such as LpH.

The World Health Organization recommends any of the following three procedures for the sterilization of all heat-resistant surgical instruments to ensure that they are not contaminated with prions:

- Immerse in 1N sodium hydroxide and place in a gravity-displacement autoclave at 121°C for 30 minutes; clean; rinse in water; and then perform routine sterilization processes.

- Immerse in 1N sodium hypochlorite (20,000 parts per million available chlorine) for 1 hour; transfer instruments to water; heat in a gravity-displacement autoclave at 121°C for 1 hour; clean; and then perform routine sterilization processes.
- Immerse in 1N sodium hydroxide or sodium hypochlorite (20,000 parts per million available chlorine) for 1 hour; remove and rinse in water, then transfer to an open pan and heat in a gravity-displacement (121°C) or in a porous-load (134°C) autoclave for 1 hour; clean; and then perform routine sterilization processes.

134 °C (273 °F) for 18 minutes in a pressurized steam autoclave has been found to be somewhat effective in deactivating the agent of disease. Ozone sterilization has been studied as a potential method for prion denaturation and deactivation. Other approaches being developed include thiourea-urea treatment, guanidinium chloride treatment, and special heat-resistant subtilisin combined with heat and detergent.

A method sufficient for sterilizing prions on one material may fail on another. Renaturation of a completely denatured prion to infectious status has not yet been achieved; however, partially denatured prions can be renatured to an infective status under certain artificial conditions.

Degradation Resistance in Nature

Overwhelming evidence shows that prions resist degradation and persist in the environment for years, and proteases do not degrade them. Experimental evidence shows that *unbound* prions degrade over time, while soil-bound prions remain at stable or increasing levels, suggesting that prions likely accumulate in the environment.

One 2015 study by US scientists found that repeated drying and wetting may render soil bound prions less infectious, although this was dependent on the soil type they were bound to.

Degradation by Living Beings

More recent studies suggest scrapie prions can be degraded by diverse cellular machinery. Inhibition of autophagy accelerates prion accumulation whereas encouragement of autophagy promotes prion clearance. The ubiquitin proteasome system appears to be able to degrade small enough aggregates. In addition, keratinase from *B. licheniformis*, alkaline serine protease from *Streptomyces sp*, subtilisin-like pernisine from *Aeropyrum pernix*, alkaline protease from *Nocardiopsis sp*, nattokinase from *B. subtilis*, engineered subtilisins from *B. lentus* and serine protease from three lichen species have been found to degrade PrP^{Sc}.

Fungi

Proteins showing prion-type behavior are also found in some fungi, which has been useful in helping to understand mammalian prions. Fungal prions do not always cause disease in their hosts. In yeast, protein refolding to the prion configuration is assisted by chaperone proteins such as Hsp104.

All known prions induce the formation of an amyloid fold, in which the protein polymerises into an aggregate consisting of tightly packed beta sheets. Amyloid aggregates are fibrils, growing at their ends, and replicate when breakage causes two growing ends to become four growing ends.

The incubation period of prion diseases is determined by the exponential growth rate associated with prion replication, which is a balance between the linear growth and the breakage of aggregates.

Fungal proteins exhibiting templated conformational change were discovered in the yeast *Saccharomyces cerevisiae* by Reed Wickner in the early 1990s. For their mechanistic similarity to mammalian prions, they were termed yeast prions. Subsequent to this, a prion has also been found in the fungus *Podospora anserina*.

These prions behave similarly to PrP, but, in general, are nontoxic to their hosts. Susan Lindquist's group at the Whitehead Institute has argued some of the fungal prions are not associated with any disease state, but may have a useful role; however, researchers at the NIH have also provided arguments suggesting that fungal prions could be considered a diseased state.

There is evidence that fungal proteins have evolved specific functions that are beneficial to the microorganism that enhance their ability to adapt to their diverse environments.

Further, within yeasts, prions can act as vectors of epigenetic inheritance, transferring traits to offspring without any genomic change.

Research into fungal prions has given strong support to the protein-only concept, since purified protein extracted from cells with a prion state has been demonstrated to convert the normal form of the protein into a misfolded form *in vitro*, and in the process, preserve the information corresponding to different strains of the prion state.

It has also shed some light on prion domains, which are regions in a protein that promote the conversion into a prion. Fungal prions have helped to suggest mechanisms of conversion that may apply to all prions, though fungal prions appear distinct from infectious mammalian prions in the lack of cofactor required for propagation. The characteristic prion domains may vary between species – e.g., characteristic fungal prion domains are not found in mammalian prions.

Table 2: Fungal Prions

Protein	Natural Host	Normal Function	Prion State	Prion Phenotype	Year Identified
Ure2p	*Saccharomyces cerevisiae*	Nitrogen catabolite repressor	[URE3]	Growth on poor nitrogen sources	1994
Sup35p	*S. cerevisiae*	Translation termination factor	[PSI+]	Increased levels of nonsense suppression	1994
HET-S	*Podospora anserina*	Regulates heterokaryon incompatibility	[Het-s]	Heterokaryon formation between incompatible strains	
Rnq1p	*S. cerevisiae*	Protein template factor	[RNQ+], [PIN+]	Promotes aggregation of other prions	
Swi1	*S. cerevisiae*	Chromatin remodeling	[SWI+]	Poor growth on some carbon sources	2008
Cyc8	*S. cerevisiae*	Transcriptional repressor	[OCT+]	Transcriptional derepression of multiple genes	2009
Mot3	*S. cerevisiae*	Nuclear transcription factor	[MOT3+]	Transcriptional derepression of anaerobic genes	2009
Sfp1	*S. cerevisiae*	Putative transcription factor	[ISP+]	Antisuppression	2010

Treatments

There are no effective treatments for prion diseases. Clinical trials in humans have not met with success and have been hampered by the rarity of prion diseases. Although some potential treatments have shown promise in the laboratory, none have been effective once the disease has commenced.

In Other Diseases

Prion-like domains have been found in a variety of other mammalian proteins. Some of these proteins have been implicated in the ontogeny of age-related neurodegenerative disorders such as amyotrophic lateral sclerosis (ALS), frontotemporal lobar degeneration with ubiquitin-positive inclusions (FTLD-U), Alzheimer's disease, Parkinson's disease, and Huntington's disease. They are also implicated in some forms of systemic amyloidosis including AA amyloidosis that develops in humans and animals with inflammatory and infectious diseases such as tuberculosis, Crohn's disease, rheumatoid arthritis, and HIV/AIDS. AA amyloidosis, like prion disease, may be transmissible. This has given rise to the 'prion paradigm', where otherwise harmless proteins can be converted to a pathogenic form by a small number of misfolded, nucleating proteins.

The definition of a prion-like domain arises from the study of fungal prions. In yeast, prionogenic proteins have a portable prion domain that is both necessary and sufficient for self-templating and protein aggregation. This has been shown by attaching the prion domain to a reporter protein, which then aggregates like a known prion. Similarly, removing the prion domain from a fungal prion protein inhibits prionogenesis. This modular view of prion behaviour has led to the hypothesis that similar prion domains are present in animal proteins, in addition to PrP. These fungal prion domains have several characteristic sequence features. They are typically enriched in asparagine, glutamine, tyrosine and glycine residues, with an asparagine bias being particularly conducive to the aggregative property of prions. Historically, prionogenesis has been seen as independent of sequence and only dependent on relative residue content. However, this has been shown to be false, with the spacing of prolines and charged residues having been shown to be critical in amyloid formation.

Bioinformatic screens have predicted that over 250 human proteins contain prion-like domains (PrLD). These domains are hypothesized to have the same transmissible, amyloidogenic properties of PrP and known fungal proteins. As in yeast, proteins involved in gene expression and RNA binding seem to be particularly enriched in PrLD's, compared to other classes of protein. In particular, 29 of the known 210 proteins with an RNA recognition motif also have a putative prion domain. Meanwhile, several of these RNA-binding proteins have been independently identified as pathogenic in cases of ALS, FTLD-U, Alzheimer's disease, and Huntington's disease.

Role in Neurodegenerative Disease

The pathogenicity of prions and proteins with prion-like domains is hypothesized to arise from their self-templating ability and the resulting exponential growth of amyloid fibrils. The presence of amyloid fibrils in patients with degenerative diseases has been well documented.

These amyloid fibrils are seen as the result of pathogenic proteins that self-propagate and form highly stable, non-functional aggregates. While this does not necessarily imply a causal relationship between amyloid and degenerative diseases, the toxicity of certain amyloid forms and the overproduction of amyloid in familial cases of degenerative disorders supports the idea that amyloid formation is generally toxic.

Specifically, aggregation of TDP-43, an RNA-binding protein, has been found in ALS/MND patients, and mutations in the genes coding for these proteins have been identified in familial cases of ALS/MND. These mutations promote the misfolding of the proteins into a prion-like conformation. The misfolded form of TDP-43 forms cytoplasmic inclusions in affected neurons, and is found depleted in the nucleus. In addition to ALS/MND and FTLD-U, TDP-43 pathology is a feature of many cases of Alzheimer's disease, Parkinson's disease and Huntington's disease. The misfolding of TDP-43 is largely directed by its prion-like domain. This domain is inherently prone to misfolding, while pathological mutations in TDP-43 have been found to increase this propensity to misfold, explaining the presence of these mutations in familial cases of ALS/MND. As in yeast, the prion-like domain of TDP-43 has been shown to be both necessary and sufficient for protein misfolding and aggregation.

Similarly, pathogenic mutations have been identified in the prion-like domains of heterogeneous nuclear riboproteins hnRNPA2B1 and hnRNPA1 in familial cases of muscle, brain, bone and motor neuron degeneration. The wild-type form of all of these proteins show a tendency to self-assemble into amyloid fibrils, while the pathogenic mutations exacerbate this behaviour and lead to excess accumulation.

Weaponization

Prions could theoretically be employed as a weaponized agent. With potential fatality rates of 100%, prions could be an effective bioweapon, sometimes called a "biochemical weapon", because a prion is a biochemical. An unfavorable aspect is prions' very long incubation periods. Persistent heavy exposure of prions to the intestine might shorten the overall onset. Another aspect of using prions in warfare is the difficulty of detection and decontamination.

5.3 MECHANISMS OF VIRAL INFECTION

The infection process of viruses is complex and varies based on the type of virus and host cell. However, it generally follows a sequence of steps, including attachment, penetration, replication, assembly, and release.

- **Attachment**: The virus attaches to the surface of the host cell through specific interactions between viral proteins and host cell receptors. This specificity determines the type of cells a virus can infect. For instance, the human immunodeficiency virus (HIV) targets immune cells that express the CD4 receptor.
- **Penetration and Uncoating**: Once attached, the virus enters the host cell. For enveloped viruses, the viral envelope fuses with the host cell membrane, releasing the viral capsid into the cytoplasm. Non-enveloped viruses may enter via endocytosis, where the host cell engulfs the virus.

The capsid is then uncoated, releasing the viral genome into the host cell.

- **Replication and Gene Expression:** The viral genome takes over the host cell's machinery to replicate and produce viral proteins. DNA viruses typically use the host's DNA polymerase to replicate their DNA, while RNA viruses rely on RNA-dependent RNA polymerase or reverse transcriptase in retroviruses to replicate their RNA.
- **Assembly and Release:** New viral particles are assembled from the replicated genome and synthesized proteins. In many cases, the virus eventually causes the host cell to lyse, releasing viral progeny into the surrounding environment. Enveloped viruses acquire their lipid envelope by budding from the host cell membrane, allowing them to exit without immediately destroying the cell.

5.4 CONTROL OF VIRAL INFECTIONS

The control of viral infections involves a multifaceted approach, including vaccination, antiviral drugs, public health measures, and prevention strategies.

- **Vaccination:** Vaccines are among the most effective tools for controlling viral diseases. They stimulate the immune system to recognize and respond to specific viruses without causing disease. Vaccines may contain live attenuated viruses, inactivated viruses, or viral proteins. For example, the measles-mumps-rubella (MMR) vaccine uses live attenuated viruses, while the flu vaccine may contain inactivated viral particles.
- **Antiviral Drugs:** Unlike antibiotics, which are ineffective against viruses, antiviral drugs are designed to inhibit specific stages of viral replication. For example, acyclovir targets DNA replication in herpesviruses, while protease inhibitors block HIV maturation. However, due to the high mutation rates of some viruses, drug resistance can emerge, complicating treatment efforts.
- **Hygiene and Sanitation:** Good hygiene practices, such as handwashing and disinfecting surfaces, play a critical role in preventing the spread of viruses. Respiratory viruses, such as influenza and SARS-CoV-2, are transmitted through droplets, so wearing masks and maintaining physical distancing are effective preventive measures.
- **Public Health Measures:** Quarantine, contact tracing, and travel restrictions are essential public health tools used to contain outbreaks. These measures are particularly important for highly infectious viruses like COVID-19, where rapid transmission can lead to widespread illness.

5.5 CHALLENGES IN CONTROLLING VIRAL INFECTIONS

Despite advances in virology and preventive medicine, controlling viral infections presents several challenges, including the high mutation rates of viruses, emergence of new pathogens, and limited access to healthcare in some regions.

- **Viral Mutation and Resistance**: Many viruses, particularly RNA viruses, mutate rapidly, leading to new strains that can evade immune responses or become resistant to antiviral drugs. This rapid evolution complicates vaccine development, as seen with the seasonal influenza virus, which requires annual vaccine updates.
- **Emerging and Re-emerging Viruses**: Viruses can jump from animal hosts to humans, leading to new outbreaks. Zoonotic viruses, like Ebola, Zika, and SARS-CoV-2, illustrate the potential of viruses to cause sudden and widespread human infections. Human activities, such as deforestation and wildlife trade, increase the risk of zoonotic transmission, highlighting the need for vigilant monitoring.
- **Global Health Disparities**: In many low-income countries, limited access to vaccines, antiviral drugs, and healthcare infrastructure hampers the control of viral diseases. Strengthening global healthcare systems and improving vaccine access are essential to managing viral outbreaks in underserved regions.

5.6 ADVANCES IN VIROLOGY AND FUTURE DIRECTIONS

The field of virology is evolving rapidly, with advances in technology and scientific research opening new avenues for understanding and controlling viral infections.

- **Genomics and Molecular Biology**: Genome sequencing has transformed virology, enabling researchers to identify viral mutations and track outbreaks in real time. Techniques such as CRISPR gene editing are being explored to develop novel antiviral therapies, potentially offering new ways to combat viral infections at the molecular level.
- **Viral Vector Vaccines and Gene Therapy**: Viral vectors, modified viruses used to deliver therapeutic genes, are revolutionizing both vaccines and gene therapy. For instance, viral vectors are used in vaccines for COVID-19, such as the Johnson & Johnson vaccine, and in treatments for genetic disorders, where they deliver corrective genes to affected cells.
- **Development of Broad-Spectrum Antivirals**: While most antivirals target specific viruses, broad-spectrum antivirals aim to inhibit a range of viruses. Researchers are investigating compounds that target

common viral mechanisms, such as RNA polymerase, offering hope for drugs that could treat multiple viral infections.

5.7 CONCLUSION

Viruses and other acellular infectious agents, including viroids and prions, are distinctive pathogens that pose significant challenges to global health. Their unique dependence on host cells for replication and their high mutation rates make them difficult to control. However, through a combination of vaccination, antiviral drugs, public health initiatives, and hygiene practices, we can mitigate their impact on human populations. Ongoing research in virology, genomics, and biotechnology promises new strategies for preventing and treating viral infections, holding the potential to reduce the burden of viral diseases worldwide. As our understanding deepens, effective management and prevention of viral outbreaks will rely on continued innovation, global collaboration, and equitable access to healthcare solutions.

CHAPTER 6
PROTOZOA

Protozoa are a diverse group of single-celled, eukaryotic microorganisms that play an essential role in ecosystems, human health, and scientific research.

Despite their simplicity, protozoa exhibit a wide range of structural diversity, life strategies, and adaptations that enable them to thrive in nearly every habitat on Earth.

6.1 CHARACTERISTICS OF PROTOZOA

Protozoa, often referred to as "animal-like" protists, are primarily unicellular and share characteristics with animals, such as mobility and heterotrophic nutrition.

They lack cell walls, differentiating them from plants and fungi, and instead rely on cell membranes for structural support and interaction with their environments.

- **Cell Structure and Organelles**: Protozoa are eukaryotic, meaning they possess a true nucleus and membrane-bound organelles, such as mitochondria, endoplasmic reticulum, and Golgi apparatus. Some protozoa contain contractile vacuoles that help regulate water balance and avoid osmotic stress, especially in freshwater environments. Their structures are adapted for various forms of movement and feeding.
- **Modes of Locomotion**: Protozoa are motile organisms and move using one of three main methods—cilia, flagella, or pseudopodia. Ciliated protozoa, like *Paramecium*, use tiny hair-like projections (cilia) to propel themselves. Flagellated protozoa, such as *Trypanosoma*, use whip-like flagella, while amoeboid protozoa, like *Amoeba proteus*, move by extending parts of their cell membrane and cytoplasm, forming pseudopodia.
- **Nutritional Strategies**: Protozoa are heterotrophic, meaning they obtain their nutrients by ingesting or absorbing organic matter. Many protozoa are phagotrophic and consume other microorganisms by engulfing them through a process called phagocytosis. Others absorb nutrients directly from their environment. Some protozoa, such as *Euglena*, are mixotrophic and can photosynthesize in the presence of light but switch to heterotrophy in darkness.

6.2 CLASSIFICATION OF PROTOZOA

The classification of protozoa has evolved over time and remains an area of scientific debate. Historically, protozoa were classified within the kingdom Protista, but modern classification divides them into several groups based on molecular data, cellular structures, and modes of locomotion. The four primary groups are Amoebozoa, Flagellates, Ciliates, and Sporozoans.

6.2.1 Amoebozoa

Amoebozoa is a major taxonomic group containing about 2,400 described species of amoeboid protists, often possessing blunt, fingerlike, lobose pseudopods and tubular mitochondrial cristae. In traditional classification schemes, Amoebozoa is usually ranked as a phylum within either the kingdom Protista or the kingdom Protozoa. In the classification favored by the International Society of Protistologists, it is retained as an unranked "supergroup" within Eukaryota. Molecular genetic analysis supports Amoebozoa as a monophyletic clade. Modern studies of eukaryotic phylogenetic trees identify it as the sister group to Opisthokonta, another major clade which contains both fungi and animals as well as several other clades comprising some 300 species of unicellular eukaryotes. Amoebozoa and Opisthokonta are sometimes grouped together in a high-level taxon, named Amorphea. Amoebozoa includes many of the best-known amoeboid organisms, such as *Chaos*, *Entamoeba*, *Pelomyxa* and the genus *Amoeba* itself. Species of Amoebozoa may be either shelled (testate) or naked, and cells may possess flagella. Free-living species are common in both salt and freshwater as well as soil, moss and leaf litter. Some live as parasites or symbionts of other organisms, and some are known to cause disease in humans and other organisms.

While the majority of amoebozoan species are unicellular, the group also includes several clades of slime molds, which have a macroscopic, multicellular stage of life during which individual amoeboid cells remain together after multiple cell division to form a macroscopic plasmodium or, in cellular slime molds, aggregate to form one.

Amoebozoa vary greatly in size. Some are only 10–20 μm in diameter, while others are among the largest protozoa. The well-known species *Amoeba proteus,* which may reach 800 μm in length, is often studied in schools and laboratories as a representative cell or model organism, partly because of its convenient size. Multinucleate amoebae like *Chaos* and *Pelomyxa* may be several millimetres in length, and some multicellular amoebozoa, such as the "dog vomit" slime mold *Fuligo septica*, can cover an area of several square meters.

Morphology

Amoebozoa is a large and diverse group, but certain features are common to many of its members. The amoebozoan cell is typically divided into a granular central mass, called endoplasm, and a clear outer layer, called ectoplasm. During locomotion, the endoplasm flows forwards and the ectoplasm runs backwards along the outside of the cell.

In motion, many amoebozoans have a clearly defined anterior and posterior and may assume a "monopodial" form, with the entire cell functioning as a single pseudopod. Large pseudopods may produce numerous clear projections called subpseudopodia (or determinate pseudopodia), which are extended to a certain length and then retracted, either for the purpose of locomotion or food intake. A cell may also form multiple indeterminate pseudopodia, through which the entire contents of the cell flow in the direction of locomotion. These are more or less tubular and are mostly filled with granular endoplasm.

The cell mass flows into a leading pseudopod, and the others ultimately retract, unless the organism changes direction.

While most amoebozoans are "naked," like the familiar *Amoeba* and *Chaos*, or covered with a loose coat of minute scales, like Cochliopodium and Korotnevella, members of the order Arcellinida form rigid shells, or tests, equipped with a single aperture through which the pseudopods emerge. Arcellinid tests may be secreted from organic materials, as in *Arcella*, or built up from collected particles cemented together, as in *Difflugia.*

In all amoebozoa, the primary mode of nutrition is phagocytosis, in which the cell surrounds potential food particles with its pseudopods, sealing them into vacuoles within which they may be digested and absorbed. Some amoebozoans have a posterior bulb called a uroid, which may serve to accumulate waste, periodically detaching from the rest of the cell.

When food is scarce, most species can form cysts, which may be carried aerially and introduce them to new environments. In slime moulds, these structures are called spores, and form on stalked structures called fruiting bodies or sporangia. Mixotrophic species living in a symbiotic relationship with microalgae of the genus Chlorella, which lives inside the cytoplasm of their host, have been found in Arcellinida and Mayorella.

The majority of Amoebozoa lack flagella and more generally do not form microtubule-supported structures except during mitosis. However, flagella do occur among the Archamoebae, and many slime moulds produce biflagellate gametes . The flagellum is generally anchored by a cone of microtubules, suggesting a close relationship to the opisthokonts. The mitochondria in amoebozoan cells characteristically have branching tubular cristae. However, among the Archamoebae, which are adapted to anoxic or microaerophilic habitats, mitochondria have been lost.

Classification

Place of Amoebozoa in the Eukaryote Tree

It appears (based on molecular genetics) that the members of Amoebozoa form a sister group to animals and fungi, diverging from this lineage after it had split from the other groups:

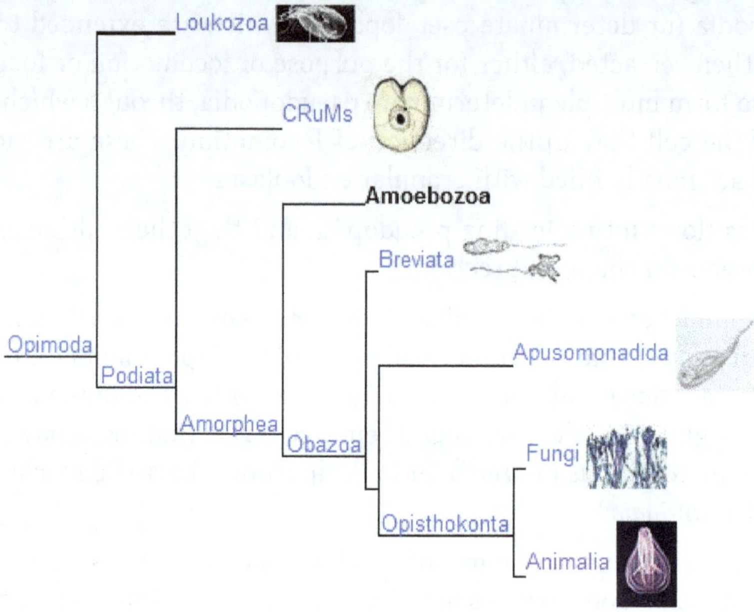

Strong similarities between Amoebozoa and Opisthokonts lead to the hypothesis that they form a distinct clade. Thomas Cavalier-Smith proposed the name "unikonts" (formally, Unikonta) for this branch, whose members were believed to have been descended from a common ancestor possessing a single emergent flagellum rooted in one basal body. However, while the close relationship between Amoebozoa and Opisthokonta is robustly supported, recent work has shown that the hypothesis of a uniciliate ancestor is probably false. In their Revised Classification of Eukaryotes (2012), Adl et al. proposed Amorphea as a more suitable name for a clade of approximately the same composition, a sister group to the Diaphoretickes. More recent work places the members of Amorphea together with the malawimonids and collodictyonids in a proposed clade called Opimoda, which comprises one of two major lineages diverging at the root of the eukaryote tree of life, the other being Diphoda.

Subphyla within Amoebozoa: Lobosa and Conosa

Traditionally all amoebozoa with lobose pseudopods were grouped together in the class Lobosea, placed with other amoeboids in the phylum Sarcodina or Rhizopoda, but these were considered to be unnatural groups. Structural and genetic studies identified the percolozoans and several archamoebae as independent groups. In phylogenies based on rRNA their representatives were separate from other amoebae, and appeared to diverge near the base of eukaryotic evolution, as did most slime molds.

However, revised trees by Cavalier-Smith and Chao in 1996 suggested that the remaining lobosans do form a monophyletic group, to which the Archamoebae and Mycetozoa were closely related, although the percolozoans were not. Subsequently, they emended the phylum Amoebozoa to include both the subphylum Lobosa and a new subphylum Conosa, comprising the Archamoebae and the Mycetozoa.

Recent molecular genetic data appear to support this primary division of the Amoebozoa into Lobosa and Conosa. The former, as defined by Cavalier-Smith and his collaborators, consists largely of the classic Lobosea: non-flagellated amoebae with blunt, lobose pseudopods (*Amoeba, Acanthamoeba, Arcella, Difflugia etc.*). The latter is made up of both amoeboid and flagellated cells, characteristically with more pointed or slightly branching subpseudopodia (Archamoebae and the Mycetozoan slime molds).

Fossil Record

Vase-shaped microfossils (VSMs) discovered around the world show that amoebozoans have existed since the Neoproterozoic Era. The fossil species *Melanocyrillium hexodiadema*, *Palaeoarcella athanata*, and *Hemisphaeriella ornata* come from rocks 750 million years old. All three VSMs share a hemispherical shape, invaginated aperture, and regular indentations, that strongly resemble modern arcellinids, which are shell-bearing amoebozoans belonging to the class Tubulinea. *P. athanata* in particular looks the same as the extant genus *Arcella*.

Meiosis

The recently available *Acanthamoeba* genome sequence revealed several orthologs of genes employed in meiosis of sexual eukaryotes. These genes included *Spo11, Mre11, Rad50, Rad51, Rad52, Mnd1, Dmc1, Msh* and *Mlh*. This finding suggests that *Acanthamoeba* is capable of some form of meiosis and may be able to undergo sexual reproduction. In sexually reproducing eukaryotes, homologous recombination (HR) ordinarily occurs during meiosis. The meiosis-specific recombinase, Dmc1, is required for efficient meiotic HR, and Dmc1 is expressed in *Entamoeba histolytica*. The purified Dmc1 from *E. histolytica* forms presynaptic filaments and catalyzes ATP-dependent homologous DNA pairing and DNA strand exchange over at least several thousand base pairs. The DNA pairing and strand exchange reactions are enhanced by the eukaryotic meiosis-specific recombination accessory factor (heterodimer) Hop2-Mnd1. These processes are central to meiotic recombination, suggesting that *E. histolytica* undergoes meiosis.

Studies of *Entamoeba invadens* found that, during the conversion from the tetraploid uninucleate trophozoite to the tetranucleate cyst, homologous recombination is enhanced. Expression of genes with functions related to the major steps of meiotic recombination also increased during encystations. These findings in *E. invadens*, combined with evidence from studies of *E. histolytica* indicate the presence of meiosis in the *Entamoeba*. A comparative genetic analysis indicated that meiotic processes are present in all major amoebozoan lineages.

Since *Amoebozoa* diverged early from the eukaryotic family tree, these results also suggest that meiosis was present early in eukaryotic evolution.

Human Health

Amoebiasis, also known as amebiasis or entamoebiasis, is an infection caused by any of the amoebozoans of the *Entamoeba* group. Symptoms are most common upon infection by *Entamoeba histolytica*. Amoebiasis can present with no, mild, or severe symptoms. Symptoms may include abdominal pain, mild diarrhoea, bloody diarrhea or severe colitis with tissue death and perforation. This last complication may cause peritonitis. People affected may develop anemia due to loss of blood.

Invasion of the intestinal lining causes amoebic bloody diarrhea or amoebic colitis. If the parasite reaches the bloodstream it can spread through the body, most frequently ending up in the liver where it causes amoebic liver abscesses. Liver abscesses can occur without previous diarrhea. Cysts of *Entamoeba* can survive for up to a month in soil or for up to 45 minutes under fingernails. It is important to differentiate between amoebiasis and bacterial colitis. The preferred diagnostic method is through faecal examination under microscope, but requires a skilled microscopist and may not be reliable when excluding infection. This method however may not be able to separate between specific types. Increased white blood cell count is present in severe cases, but not in mild ones. The most accurate test is for antibodies in the blood, but it may remain positive following treatment.

Prevention of amoebiasis is by separating food and water from faeces and by proper sanitation measures. There is no vaccine. There are two treatment options depending on the location of the infection. Amoebiasis in tissues is treated with either metronidazole, tinidazole, nitazoxanide, dehydroemetine or chloroquine, while luminal infection is treated with diloxanide furoate or iodoquinoline. For treatment to be effective against all stages of the amoeba may require a combination of medications. Infections without symptoms do not require treatment but infected individuals can spread the parasite to others and treatment can be considered. Treatment of other *Entamoeba* infections apart from *E. histolytica* is not needed.

Amoebiasis is present all over the world. About 480 million people are infected with what appears to be *E. histolytica* and these result in the death of between

40,000–110,000 people every year. Most infections are now ascribed to *E. dispar. E. dispar* is more common in certain areas and symptomatic cases may be fewer than previously reported. The first case of amoebiasis was documented in 1875 and in 1891 the disease was described in detail, resulting in the terms *amoebic dysentery* and *amoebic liver abscess.* Further evidence from the Philippines in 1913 found that upon ingesting cysts of *E. histolytica* volunteers developed the disease. It has been known since 1897 that at least one non-disease-causing species of *Entamoeba* existed (*Entamoeba coli*), but it was first formally recognized by the WHO in 1997 that *E. histolytica* was two species, despite this having first been proposed in 1925. In addition to the now-recognized *E. dispar* evidence shows there are at least two other species of *Entamoeba* that look the same in humans - *E. moshkovskii* and *Entamoeba bangladeshi.* The reason these species haven't been differentiated until recently is because of the reliance on appearance.

6.2.2 Flagellates (Mastigophora)

A flagellate is a cell or organism with one or more whip-like appendages called flagella. The word *flagellate* also describes a particular construction (or level of organization) characteristic of many prokaryotes and eukaryotes and their means of motion. The term presently does not imply any specific relationship or classification of the organisms that possess flagella. However, several derivations of the term "flagellate" (such as "dinoflagellate" and "choanoflagellata") are more formally characterized.

Flagella in eukaryotes are supported by microtubules in a characteristic arrangement, with nine fused pairs surrounding two central singlets. These arise from a basal body. In some flagellates, flagella direct food into a cytostome or mouth, where food is ingested. Flagella role in classifying eukaryotes.

Among protoctists and microscopic animals, a flagellate is an organism with one or more flagella. Some cells in other animals may be flagellate, for instance the spermatozoa of most animal phyla. Flowering plants do not produce flagellate cells, but ferns, mosses, green algae, and some gymnosperms and closely related plants do so. Likewise, most fungi do not produce cells with flagellae, but the primitive fungal chytrids do. Many protists take the form of single-celled flagellates.

Flagella are generally used for propulsion. They may also be used to create a current that brings in food. In most such organisms, one or more flagella are located at or near the anterior of the cell (e.g., *Euglena*). Often there is one directed forwards and one trailing behind. Many parasites that affect human health or economy are flagellates in at least one stage of life cycle, such as *Naegleria*, *Trichomonas* and *Plasmodium*. Flagellates are the major consumers of primary and secondary production in aquatic ecosystems - consuming bacteria and other protists.

6.2.3 Ciliates (Ciliophora)

The ciliates are a group of alveolates characterized by the presence of hair-like organelles called cilia, which are identical in structure to eukaryotic flagella, but are in general shorter and present in much larger numbers, with a different undulating pattern than flagella. Cilia occur in all members of the group (although the peculiar Suctoria only have them for part of their life cycle) and are variously used in swimming, crawling, attachment, feeding, and sensation.

Ciliates are an important group of protists, common almost anywhere there is water—in lakes, ponds, oceans, rivers, and soils, including anoxic and oxygen-depleted habitats. About 4,500 unique free-living species have been described, and the potential number of extant species is estimated at 27,000–40,000. Included in this number are many ectosymbiotic and endosymbiotic species, as well as some obligate and opportunistic parasites. Ciliate species range in size from as little as 10 µm in some colpodeans to as much as 4 mm in length in some geleiids, and include some of the most morphologically complex protozoans.

In most systems of taxonomy, "Ciliophora" is ranked as a phylum under any of several kingdoms, including Chromista, Protista or Protozoa. In some older systems of classification, such as the influential taxonomic works of Alfred Kahl, ciliated protozoa are placed within the class "Ciliata" (a term which can also refer to a genus of fish). In the taxonomic scheme endorsed by the International Society of Protistologists, which eliminates formal rank designations such as "phylum" and "class", "Ciliophora" is an unranked taxon within Alveolata.

Cell Structure

Nuclei

Unlike most other eukaryotes, ciliates have two different sorts of nuclei: a tiny, diploid micronucleus (the "generative nucleus", which carries the germline of the cell), and a large, ampliploid macronucleus (the "vegetative nucleus", which takes care of general cell regulation, expressing the phenotype of the organism). The latter is generated from the micronucleus by amplification of the genome and heavy editing. The micronucleus passes its genetic material to offspring, but does not express its genes. The macronucleus provides the small nuclear RNA for vegetative growth.

Division of the macronucleus occurs in most ciliate species, apart from those in class Karyorelictea, whose macronuclei are replaced every time the cell divides. Macronuclear division is accomplished by amitosis, and the segregation of the chromosomes occurs by a process whose mechanism is unknown. After a certain number of generations (200–350, in *Paramecium aurelia*, and as many as 1,500 in *Tetrahymena*) the cell shows signs of aging, and the macronuclei must be regenerated from the micronuclei. Usually, this occurs following *conjugation*, after which a new macronucleus is generated from the post-conjugal micronucleus.

Cytoplasm

Food vacuoles are formed through phagocytosis and typically follow a particular path through the cell as their contents are digested and broken down by lysosomes so the substances the vacuole contains are then small enough to diffuse through the membrane of the food vacuole into the cell. Anything left in the food vacuole by the time it reaches the cytoproct (anal pore) is discharged by exocytosis. Most ciliates also have one or more prominent contractile vacuoles, which collect water and expel it from the cell to maintain osmotic pressure, or in some function to maintain ionic balance. In some genera, such as *Paramecium*, these have a distinctive star shape, with each point being a collecting tube.

Specialized Structures in Ciliates

Mostly, body cilia are arranged in *mono-* and *dikinetids*, which respectively include one and two kinetosomes (basal bodies), each of which may support a cilium. These are arranged into rows called *kineties*, which run from the anterior to posterior of the cell. The body and oral kinetids make up the *infraciliature*, an organization unique to the ciliates and important in their classification, and include various fibrils and microtubules involved in coordinating the cilia. In some forms there are also body polykinetids, for instance, among the spirotrichs where they generally form bristles called cirri.

The infraciliature is one of the main components of the cell cortex. Others are the *alveoli*, small vesicles under the cell membrane that are packed against it to form a pellicle maintaining the cell's shape, which varies from flexible and contractile to rigid. Numerous mitochondria and extrusomes are also generally present. The presence of alveoli, the structure of the cilia, the form of mitosis and various other details indicate a close relationship between the ciliates, Apicomplexa, and dinoflagellates. These superficially dissimilar groups make up the alveolates.

Feeding

Most ciliates are heterotrophs, feeding on smaller organisms, such as bacteria and algae, and detritus swept into the oral groove (mouth) by modified oral cilia. This usually includes a series of membranelles to the left of the mouth and a paroral membrane to its right, both of which arise from *polykinetids*, groups of many cilia together with associated structures. The food is moved by the cilia through the mouth pore into the gullet, which forms food vacuoles.

Many species are also mixotrophic, combining phagotrophy and phototrophy through kleptoplasty or symbiosis with photosynthetic microbes. The ciliate *Halteria* has been observed to feed on chloroviruses. Feeding techniques vary considerably, however. Some ciliates are mouthless and feed by absorption (osmotrophy), while others are predatory and feed on other protozoa and in particular on other ciliates. Some ciliates parasitize animals, although only one species, *Balantidium coli*, is known to cause disease in humans.

Reproduction and Sexual Phenomena

Reproduction

Ciliates reproduce asexually, by various kinds of fission. During fission, the micronucleus undergoes mitosis and the macronucleus elongates and undergoes amitosis (except among the Karyorelictean ciliates, whose macronuclei do not divide). The cell then divides in two, and each new cell obtains a copy of the micronucleus and the macronucleus.

Typically, the cell is divided transversally, with the anterior half of the ciliate (the *proter*) forming one new organism, and the posterior half (the *opisthe*) forming another. However, other types of fission occur in some ciliate groups. These include *budding* (the emergence of small ciliated offspring, or "swarmers", from the body of a mature parent); *strobilation* (multiple divisions along the cell body, producing a chain of new organisms); and *palintomy* (multiple fissions, usually within a cyst).

Fission may occur spontaneously, as part of the vegetative cell cycle. Alternatively, it may proceed as a result of self-fertilization (autogamy), or it may follow conjugation, a sexual phenomenon in which ciliates of compatible mating types exchange genetic material. While conjugation is sometimes described as a form of reproduction, it is not directly connected with reproductive processes, and does not directly result in an increase in the number of individual ciliates or their progeny.

Conjugation

Ciliate conjugation is a sexual phenomenon that results in genetic recombination and nuclear reorganization within the cell. During conjugation, two ciliates of a compatible mating type form a bridge between their cytoplasms. The micronuclei undergo meiosis, the macronuclei disappear, and haploid micronuclei are exchanged over the bridge. In some ciliates (peritrichs, chonotrichs and some suctorians), conjugating cells become permanently fused, and one conjugant is absorbed by the other. In most ciliate groups, however, the cells separate after conjugation, and both form new macronuclei from their micronuclei. Conjugation and autogamy are always followed by fission.

In many ciliates, such as *Paramecium*, conjugating partners (gamonts) are similar or indistinguishable in size and shape. This is referred to as "isogamontic" conjugation.

In some groups, partners are different in size and shape. This is referred to as "anisogamontic" conjugation. In sessile peritrichs, for instance, one sexual partner (the microconjugant) is small and mobile, while the other (macroconjugant) is large and sessile.

Stages of Conjugation

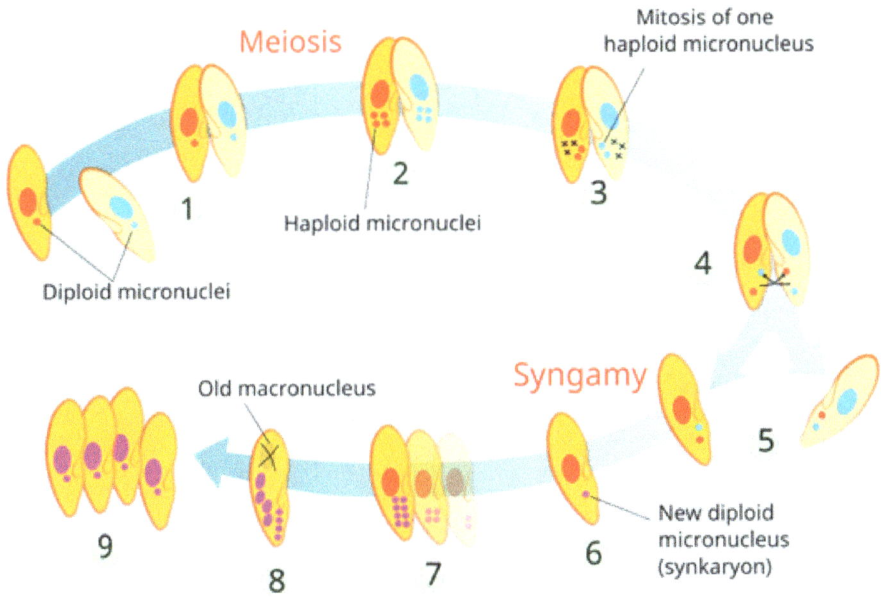

Fig 1: Stages of Conjugation in *Paramecium Caudatum*

In *Paramecium caudatum*, the stages of conjugation are as follows:

1. Compatible mating strains meet and partly fuse.
2. The micronuclei undergo meiosis, producing four haploid micronuclei per cell.
3. Three of these micronuclei disintegrate. The fourth undergoes mitosis.
4. The two cells exchange a micronucleus.
5. The cells then separate.
6. The micronuclei in each cell fuse, forming a diploid micronucleus.
7. Mitosis occurs three times, giving rise to eight micronuclei.
8. Four of the new micronuclei transform into macronuclei, and the old macronucleus disintegrates.
9. Binary fission occurs twice, yielding four identical daughter cells.

DNA Rearrangements (Gene Scrambling)

Ciliates contain two types of nuclei: somatic "macronucleus" and the germline "micronucleus". Only the DNA in the micronucleus is passed on during sexual reproduction (conjugation). On the other hand, only the DNA in the macronucleus is actively expressed and results in the phenotype of the organism. Macronuclear DNA is derived from micronuclear DNA by amazingly extensive DNA rearrangement and amplification.

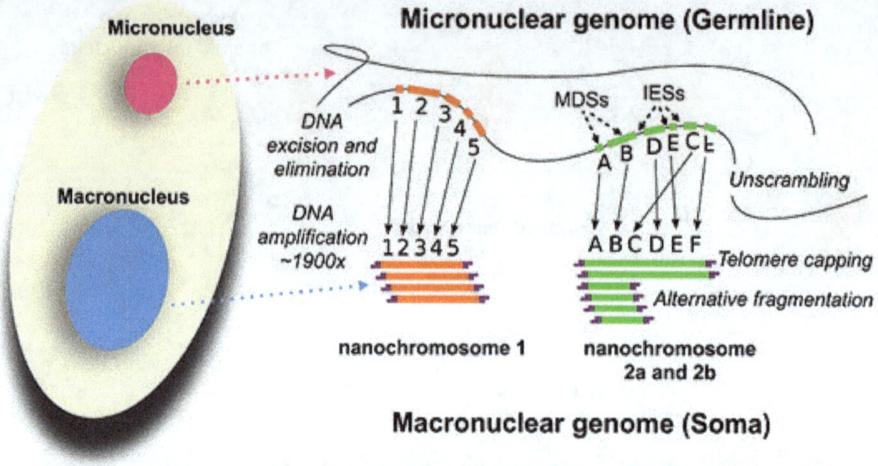

Fig 2: Development of the *Oxytricha* Macronuclear Genome from Micronuclear Genome

The macronucleus begins as a copy of the micronucleus. The micronuclear chromosomes are fragmented into many smaller pieces and amplified to give many copies. The resulting macronuclear chromosomes often contain only a single gene. In *Tetrahymena*, the micronucleus has 10 chromosomes (five per haploid genome), while the macronucleus has over 20,000 chromosomes.

In addition, the micronuclear genes are interrupted by numerous "internal eliminated sequences" (IESs). During development of the macronucleus, IESs are deleted and the remaining gene segments, macronuclear destined sequences (MDSs), are spliced together to give the operational gene.

Tetrahymena has about 6,000 IESs and about 15% of micronuclear DNA is eliminated during this process. The process is guided by small RNAs and epigenetic chromatin marks.

In spirotrich ciliates (such as *Oxytricha*), the process is even more complex due to "gene scrambling": the MDSs in the micronucleus are often in different order and orientation from that in the macronuclear gene, and so in addition to deletion, DNA inversion and translocation are required for "unscrambling". This process is guided by long RNAs derived from the parental macronucleus. More than 95% of micronuclear DNA is eliminated during spirotrich macronuclear development.

Aging

In clonal populations of *Paramecium*, aging occurs over successive generations leading to a gradual loss of vitality, unless the cell line is revitalized by conjugation or autogamy. In *Paramecium tetraurelia*, the clonally aging line loses vitality and expires after about 200 fissions, if the cell line is not rejuvenated by conjugation or self-fertilization.

The basis for clonal aging was clarified by the transplantation experiments of Aufderheide in 1986 who demonstrated that the macronucleus, rather than the cytoplasm, is responsible for clonal aging. Additional experiments by Smith-Sonneborn, Holmes and Holmes, and Gilley and Blackburn demonstrated that, during clonal aging, DNA damage increases dramatically. Thus, DNA damage appears to be the cause of aging in *P. tetraurelia*.

Fossil Record

Until recently, the oldest ciliate fossils known were tintinnids from the Ordovician period. In 2007, Li *et al.* published a description of fossil ciliates from the Doushantuo Formation, about 580 million years ago, in the Ediacaran period. These included two types of tintinnids and a possible ancestral suctorian. A fossil *Vorticella* has been discovered inside a leech cocoon from the Triassic period, about 200 million years ago.

6.2.4 Sporozoans (Apicomplexa)

The Apicomplexa (also called Apicomplexia; single: apicomplexan) are organisms of a large phylum of mainly parasitic alveolates. Most possess a unique form of organelle structure that comprises a type of non-photosynthetic plastid called an apicoplast—with an apical complex membrane. The organelle's apical shape is an adaptation that the apicomplexan applies in penetrating a host cell.

The Apicomplexa are unicellular and spore-forming. Most are obligate endoparasites of animals, except *Nephromyces*, a symbiont in marine animals, originally classified as a chytrid fungus, and the Chromerida, some of which are photosynthetic partners of corals. Motile structures such as flagella or pseudopods are present only in certain gamete stages.

The phylum Apicomplexa contains all eukaryotes with a group of structures and organelles collectively termed the apical complex. This complex consists of structural components and secretory organelles required for invasion of host cells during the parasitic stages of the Apicomplexan life cycle. Apicomplexa have complex life cycles, involving several stages and typically undergoing both asexual and sexual replication. All Apicomplexa are obligate parasites for some portion of their life cycle, with some parasitizing two separate hosts for their asexual and sexual stages.

Besides the conserved apical complex, Apicomplexa are morphologically diverse. Different organisms within Apicomplexa, as well as different life stages for a given apicomplexan, can vary substantially in size, shape, and subcellular structure. Like other eukaryotes, Apicomplexa have a nucleus, endoplasmic reticulum and Golgi complex. Apicomplexa generally have a single mitochondrion, as well as another endosymbiont-derived organelle called the apicoplast which maintains a separate 35 kilobase circular genome (with the exception of *Cryptosporidium* species and *Gregarina niphandrodes* which lack an apicoplast).

All members of this phylum have an infectious stage—the sporozoite—which possesses three distinct structures in an apical complex. The apical complex consists of a set of spirally arranged microtubules (the conoid), a secretory body (the rhoptry) and one or more polar rings. Additional slender electron-dense secretory bodies (micronemes) surrounded by one or two polar rings may also be present. This structure gives the phylum its name. A further group of spherical organelles is distributed throughout the cell rather than being localized at the apical complex and are known as the dense granules. These typically have a mean diameter around 0.7 μm. Secretion of the dense-granule content takes place after parasite invasion and localization within the parasitophorous vacuole and persists for several minutes.

- Flagella are found only in the motile gamete. These are posteriorly directed and vary in number (usually one to three).
- Basal bodies are present. Although hemosporidians and piroplasmids have normal triplets of microtubules in their basal bodies, coccidians and gregarines have nine singlets.
- The mitochondria have tubular cristae.
- Centrioles, chloroplasts, ejectile organelles, and inclusions are absent.
- The cell is surrounded by a pellicle of three membrane layers (the alveolar structure) penetrated by micropores.

Life Cycle

Most members have a complex lifecycle, involving both asexual and sexual reproduction. Typically, a host is infected via an active invasion by the parasites (similar to entosis), which divide to produce sporozoites that enter its cells. Eventually, the cells burst, releasing merozoites, which infect new cells. This may occur several times, until gamonts are produced, forming gametes that fuse to create new cysts. Many variations occur on this basic pattern, however, and many Apicomplexa have more than one host. The apical complex includes vesicles called rhoptries and micronemes, which open at the anterior of the cell. These secrete enzymes that allow the parasite to enter other cells. The tip is surrounded by a band of microtubules, called the polar ring, and among the Conoidasida is also a funnel of tubulin proteins called the conoid. Over the rest of the cell, except for a diminished mouth called the micropore, the membrane is supported by vesicles called alveoli, forming a semirigid pellicle.

The presence of alveoli and other traits place the Apicomplexa among a group called the alveolates. Several related flagellates, such as *Perkinsus* and *Colpodella*, have structures similar to the polar ring and were formerly included here, but most appear to be closer relatives of the dinoflagellates. They are probably similar to the common ancestor of the two groups.

Another similarity is that many apicomplexan cells contain a single plastid, called the apicoplast, surrounded by either three or four membranes. Its functions are thought to include tasks such as lipid and heme biosynthesis, and it appears to be necessary for survival. In general, plastids are considered to have a common origin with the chloroplasts of dinoflagellates, and evidence points to an origin from red algae rather than green.

Subgroups

Within this phylum are four groups — coccidians, gregarines, haemosporidians (or haematozoans, including in addition piroplasms), and marosporidians. The coccidians and haematozoans appear to be relatively closely related. *Perkinsus* , while once considered a member of the Apicomplexa, has been moved to a new phylum — Perkinsozoa.

Gregarines

The gregarines are generally parasites of annelids, arthropods, and molluscs. They are often found in the guts of their hosts, but may invade the other tissues. In the typical gregarine lifecycle, a trophozoite develops within a host cell into a schizont. This then divides into a number of merozoites by schizogony. The merozoites are released by lysing the host cell, which in turn invade other cells. At some point in the apicomplexan lifecycle, gametocytes are formed. These are released by lysis of the host cells, which group together. Each gametocyte forms multiple gametes. The gametes fuse with another to form oocysts. The oocysts leave the host to be taken up by a new host.

Coccidians

In general, coccidians are parasites of vertebrates. Like gregarines, they are commonly parasites of the epithelial cells of the gut, but may infect other tissues. The coccidian lifecycle involves merogony, gametogony, and sporogony. While similar to that of the gregarines it differs in zygote formation. Some trophozoites enlarge and become macrogamete, whereas others divide repeatedly to form microgametes (anisogamy). The microgametes are motile and must reach the macrogamete to fertilize it. The fertilized macrogamete forms a zygote that in its turn forms an oocyst that is normally released from the body. Syzygy, when it occurs, involves markedly anisogamous gametes. The lifecycle is typically haploid, with the only diploid stage occurring in the zygote, which is normally short-lived.

The main difference between the coccidians and the gregarines is in the gamonts. In the coccidia, these are small, intracellular, and without epimerites or mucrons. In the gregarines, these are large, extracellular, and possess epimerites or mucrons. A second difference between the coccidia and the gregarines also lies in the gamonts. In the coccidians, a single gamont becomes a macrogametocyte, whereas in the gregarines, the gamonts give rise to multiple gametocytes.

Haemosporidia

The Haemosporidia have more complex lifecycles that alternate between an arthropod and a vertebrate host. The trophozoite parasitises erythrocytes or other tissues in the vertebrate host. Microgametes and macrogametes are always found in the blood. The gametes are taken up by the insect vector during a blood meal. The microgametes migrate within the gut of the insect vector and fuse with the macrogametes. The fertilized macrogamete now becomes an ookinete, which penetrates the body of the vector. The ookinete then transforms into an oocyst and divides initially by meiosis and then by mitosis (haplontic lifecycle) to give rise to the sporozoites. The sporozoites escape from the oocyst and migrate within the body of the vector to the salivary glands where they are injected into the new vertebrate host when the insect vector feeds again.

Marosporida

The class Marosporida Mathur, Kristmundsson, Gestal, Freeman, and Keeling 2020 is a newly recognized lineage of apicomplexans that is sister to the Coccidia and Hematozoa. It is defined as a phylogenetic clade containing *Aggregata octopiana* Frenzel 1885, *Merocystis kathae* Dakin, 1911 (both Aggregatidae, originally coccidians), *Rhytidocystis* sp. 1 and *Rhytidocystis* sp. 2 Janouškovec et al. 2019 (Rhytidocystidae Levine, 1979, originally coccidians, Agamococcidiorida), and *Margolisiella islandica* Kristmundsson et al. 2011 (closely related to Rhytidocystidae). Marosporida infect marine invertebrates. Members of this clade retain plastid genomes and the canonical apicomplexan plastid metabolism. However, marosporidians have the most reduced apicoplast genomes sequenced to date, lack canonical plastidial RNA polymerase and so provide new insights into reductive organelle evolution.

Ecology and Distribution

Many of the apicomplexan parasites are important pathogens of humans and domestic animals. In contrast to bacterial pathogens, these apicomplexan parasites are eukaryotic and share many metabolic pathways with their animal hosts. This makes therapeutic target development extremely difficult – a drug that harms an apicomplexan parasite is also likely to harm its human host. At present, no effective vaccines are available for most diseases caused by these parasites. Biomedical research on these parasites is challenging because it is often difficult, if not impossible, to maintain live parasite cultures in the laboratory and to genetically manipulate these organisms. In recent years, several of the apicomplexan species have been selected for genome sequencing.

The availability of genome sequences provides a new opportunity for scientists to learn more about the evolution and biochemical capacity of these parasites. The predominant source of this genomic information is the EuPathDB family of websites, which currently provides specialised services for *Plasmodium* species (PlasmoDB), coccidians (ToxoDB), piroplasms (PiroplasmaDB), and *Cryptosporidium* species (CryptoDB). One possible target for drugs is the plastid, and in fact existing drugs such as tetracyclines, which are effective against apicomplexans, seem to operate against the plastid.

Many Coccidiomorpha have an intermediate host, as well as a primary host, and the evolution of hosts proceeded in different ways and at different times in these groups. For some coccidiomorphs, the original host has become the intermediate host, whereas in others it has become the definitive host. In the genera *Aggregata*, *Atoxoplasma*, *Cystoisospora*, *Schellackia*, and *Toxoplasma*, the original is now definitive, whereas in *Akiba*, *Babesiosoma*, *Babesia*, *Haemogregarina*, *Haemoproteus*, *Hepatozoon*, *Karyolysus*, *Leucocytozoon*, *Plasmodium*, *Sarcocystis*, and *Theileria*, the original hosts are now intermediate.

Similar strategies to increase the likelihood of transmission have evolved in multiple genera. Polyenergid oocysts and tissue cysts are found in representatives of the orders Protococcidiorida and Eimeriida. Hypnozoites are found in *Karyolysus lacerate* and most species of *Plasmodium*; transovarial transmission of parasites occurs in lifecycles of *Karyolysus* and *Babesia*.

Horizontal gene transfer appears to have occurred early on in this phylum's evolution with the transfer of a histone H4 lysine 20 (H4K20) modifier, KMT5A (Set8), from an animal host to the ancestor of apicomplexans. A second gene—H3K36 methyltransferase (Ashr3 in plants)—may have also been horizontally transferred.

Blood-Borne Genera

Within the Apicomplexa are three suborders of parasites:

- **Suborder Adeleorina:** Eight genera.
- **Suborder Laveraniina (Formerly Haemosporina):** All genera in this suborder.
- **Suborder Eimeriorina:** Two genera (*Lankesterella* and *Schellackia*).

Within the Adelorina are species that infect invertebrates and others that infect vertebrates. The Eimeriorina—the largest suborder in this phylum—the lifecycle involves both sexual and asexual stages.

The asexual stages reproduce by schizogony. The male gametocyte produces a large number of gametes and the zygote gives rise to an oocyst, which is the infective stage. The majority are monoxenous (infect one host only), but a few are heteroxenous (lifecycle involves two or more hosts).

The number of families in this later suborder is debated, with the number of families being between one and 20 depending on the authority and the number of genera being between 19 and 25.

6.3 LIFE CYCLES OF PROTOZOA

Protozoa exhibit a wide range of life cycles, from simple binary fission to complex cycles involving multiple hosts and stages. Understanding these life cycles is crucial for studying their biology and managing diseases they cause.

- **Asexual Reproduction:** Most protozoa reproduce asexually, primarily through binary fission, where a single cell divides into two identical daughter cells. Some protozoa also reproduce through budding, where a smaller cell forms on the parent cell and eventually separates. Ciliates undergo a unique type of asexual reproduction called conjugation, in which they exchange genetic material to increase genetic diversity.
- **Sexual Reproduction:** Some protozoa, particularly sporozoans, exhibit sexual reproduction during certain stages of their life cycle. For example, *Plasmodium* undergoes sexual reproduction in the mosquito host and asexual reproduction in the human host. Sexual reproduction in protozoa often increases genetic variation, aiding in adaptation to changing environments or hosts.
- **Encystment:** To survive harsh conditions, some protozoa form cysts, which are dormant forms with a protective outer layer. Encystment helps protozoa withstand extreme temperatures, lack of food, and desiccation. This ability to encyst is particularly important for parasitic protozoa, allowing them to survive outside their host and facilitating transmission.

6.4 ECOLOGICAL ROLES OF PROTOZOA

Protozoa play an essential role in ecosystems, contributing to nutrient cycling, controlling bacterial populations, and forming the base of aquatic food chains.

- **Nutrient Cycling:** Protozoa help decompose organic matter in soil and aquatic environments, releasing nutrients back into the ecosystem. By breaking down plant and animal material, protozoa contribute to the nutrient cycle, supporting plant growth and microbial diversity.

- **Predators of Microbes:** Protozoa are significant predators of bacteria and algae, helping control microbial populations. This grazing behavior prevents the overgrowth of certain bacterial species and maintains microbial diversity, which is crucial for ecosystem stability.
- **Symbiotic Relationships:** Some protozoa have symbiotic relationships with other organisms. For instance, *Trichonympha* resides in the guts of termites, where it assists in digesting cellulose. Similarly, some ciliates have symbiotic algae that photosynthesize, providing food for the protozoa in exchange for protection.

6.5 PROTOZOA AND HUMAN HEALTH

While many protozoa are harmless or beneficial, certain species cause severe diseases in humans, posing major health challenges globally.

- **Parasitic Protozoa:** Protozoan parasites, such as *Plasmodium* (malaria), *Trypanosoma* (sleeping sickness and Chagas disease), *Leishmania* (leishmaniasis), and *Giardia lamblia* (giardiasis), are responsible for millions of infections and deaths each year. These diseases are often transmitted through insect vectors, such as mosquitoes or sandflies, or contaminated water.
- **Protozoan Infections and Impact:** Protozoan infections can range from mild to life-threatening. Malaria, for instance, is a significant cause of morbidity and mortality, especially in tropical regions. These infections often present challenges in treatment due to drug resistance and require coordinated public health efforts for control and prevention.
- **Research in Medicine:** Protozoa serve as valuable research models in understanding cell biology, genetics, and evolution. The study of protozoa has led to discoveries in cellular processes and the development of medications for protozoan diseases. For example, research on *Plasmodium* has significantly contributed to our knowledge of parasitic life cycles and drug resistance.

6.6 CONTROL AND MANAGEMENT OF PROTOZOAN DISEASES

Controlling protozoan infections requires a combination of preventive measures, medical treatments, and public health initiatives.

- **Preventive Measures:** Vector control, such as insecticide-treated bed nets and indoor spraying, is crucial for diseases transmitted by insect vectors. Access to clean water and improved sanitation also helps prevent waterborne protozoan infections like giardiasis.

- **Medications and Treatment**: Treating protozoan infections involves antiparasitic drugs, though resistance is a growing problem. For instance, antimalarial drugs like chloroquine and artemisinin are used to treat malaria, while metronidazole is effective against amoebic dysentery. Continuous research is needed to develop new drugs and combat resistance.
- **Public Health Interventions**: Efforts to educate at-risk populations and improve healthcare access are essential in reducing the spread and impact of protozoan diseases. Vaccination research is ongoing for certain protozoan diseases, such as malaria, with promising developments in recent years.

6.7 CONCLUSION

Protozoa are fascinating and diverse microorganisms with far-reaching effects on ecosystems and human health. Their role in nutrient cycling, microbial population control, and symbiosis highlights their ecological importance. However, certain protozoa also cause significant diseases that impact global health, particularly in tropical and low-resource regions. By understanding the characteristics, life cycles, and ecological roles of protozoa, researchers can better manage their benefits and mitigate their harmful effects. Continued research on protozoa offers valuable insights into evolutionary biology, cell function, and potential treatments for protozoan infections, contributing to advancements in both science and public health.

PART 3:

PATHOGENS AND HOST RESPONSE

CHAPTER 7

PRINCIPLES OF MICROBIAL PATHOGENICITY AND EPIDEMIOLOGY

Microbial pathogenicity and epidemiology are core aspects of understanding infectious diseases and their impact on public health. Pathogenicity refers to the ability of microorganisms—such as bacteria, viruses, fungi, and protozoa—to cause disease, while epidemiology studies how these diseases spread, their causes, and the strategies for prevention and control. Together, these principles are essential for devising methods to control infections and mitigate public health risks.

7.1 MICROBIAL PATHOGENICITY: AN OVERVIEW

Microbial pathogenicity is the ability of a microorganism to cause disease in a host. Not all microbes are pathogenic; many are harmless or even beneficial to humans. However, certain microbes possess characteristics that enable them to invade hosts, evade immune responses, and cause disease. These organisms are known as pathogens, and their capacity to cause disease depends on factors such as virulence, mode of transmission, and the susceptibility of the host.

- **Definition and Types of Pathogens:** Pathogens can be categorized into four main types: bacteria, viruses, fungi, and protozoa. Each type of pathogen has unique mechanisms of infecting hosts and causing disease. For instance, bacteria produce toxins that can harm host cells, while viruses replicate within host cells, disrupting cellular function.
- **Pathogenicity vs. Virulence:** While pathogenicity refers to the ability to cause disease, virulence describes the degree of pathogenicity or the severity of disease caused by a pathogen. Highly virulent pathogens cause more severe diseases than less virulent ones. Virulence factors are the specific properties of a pathogen that contribute to its ability to cause disease, such as toxins, adhesins, and enzymes that break down host tissues.

7.2 MECHANISMS OF PATHOGENICITY

The process of infection typically involves several stages, from entry into the host to the expression of disease symptoms. Pathogens use a variety of mechanisms to bypass host defenses, replicate, and cause harm to the host. Understanding these mechanisms helps researchers develop treatments and preventive measures.

- **Attachment and Colonization**: For an infection to begin, the pathogen must attach to host cells, a process known as adhesion. Pathogens have specialized structures, such as pili and adhesins, that allow them to bind to specific receptors on host cells. This attachment is essential for colonization, where the pathogen establishes itself in the host tissues.
- **Invasion and Spread**: After colonization, some pathogens invade host tissues to spread and access nutrients. They may produce enzymes, such as collagenase and hyaluronidase, that break down host tissue barriers, allowing them to penetrate deeper. Pathogens that invade tissues can cause localized infections, like abscesses, or systemic infections that spread throughout the body.
- **Evasion of Host Immune Responses**: Pathogens have developed strategies to evade the host immune system, ensuring their survival and proliferation. Some bacteria produce capsules that protect them from being engulfed by immune cells, while others, like certain viruses, alter their surface proteins to avoid immune detection. Additionally, some pathogens hide within host cells, effectively avoiding the immune response.
- **Production of Toxins**: Many pathogenic bacteria produce toxins that directly harm the host. Toxins are classified into two main types: exotoxins and endotoxins. Exotoxins are proteins secreted by bacteria that target specific cells, causing damage. For instance, the botulinum toxin causes paralysis by blocking nerve signals. Endotoxins, on the other hand, are components of the outer membrane of Gram-negative bacteria and trigger inflammation when released into the host.

7.3 HOST FACTORS INFLUENCING PATHOGENICITY

The ability of a pathogen to cause disease also depends on host factors, including the immune system, genetics, and overall health. Host susceptibility varies widely, and factors such as age, underlying health conditions, and lifestyle choices can increase vulnerability to infections.

- **Immune System Function**: A strong immune system can often prevent infections or limit their severity. However, individuals with weakened immune systems—such as the elderly, infants, or immunocompromised individuals—are at higher risk of infection and may experience more severe disease.

- **Genetic Factors:** Genetic differences among individuals can influence susceptibility to infections. For instance, people with certain genetic mutations may lack receptors that specific pathogens need to attach to, reducing their risk of infection. Genetic predispositions also affect immune response and disease outcomes.
- **Environmental and Lifestyle Factors:** Lifestyle factors, such as diet, stress, and exposure to environmental toxins, can affect a person's susceptibility to infection. Living conditions, such as overcrowded environments, increase the likelihood of pathogen transmission, as do behaviors like smoking, which can compromise respiratory health.

7.4 PRINCIPLES OF EPIDEMIOLOGY

Epidemiology is the study of how diseases spread, the factors contributing to their spread, and methods for controlling outbreaks. Epidemiologists analyze patterns of disease occurrence and develop strategies to prevent and manage infections.

- **Disease Transmission:** Understanding how diseases spread is essential in epidemiology. Infectious diseases can be transmitted through direct contact, such as physical contact or respiratory droplets, or indirectly, through vectors like mosquitoes or contaminated surfaces. Modes of transmission are classified as horizontal (from person to person) or vertical (from parent to offspring).
- **Incidence and Prevalence:** Epidemiologists study the frequency of diseases using two key metrics: incidence and prevalence. Incidence refers to the number of new cases of a disease in a specific period, while prevalence indicates the total number of cases at a given time. These metrics help identify trends, track outbreaks, and evaluate public health interventions.
- **Epidemic, Endemic, and Pandemic Diseases:** Diseases are classified based on their geographic spread. Endemic diseases are consistently present in a specific region, such as malaria in certain tropical areas. Epidemics occur when there is a sudden increase in disease cases in a particular region, while pandemics are global epidemics affecting multiple countries, such as the COVID-19 pandemic.

7.5 CONTROL AND PREVENTION STRATEGIES

Controlling infectious diseases requires a combination of strategies, including surveillance, vaccination, hygiene measures, and education. Epidemiologists play a key role in designing and implementing these interventions to reduce disease transmission and improve public health outcomes.

- **Vaccination Programs**: Vaccination is one of the most effective tools for preventing infectious diseases. By stimulating the immune system to produce a response against specific pathogens, vaccines help build immunity within populations. Mass vaccination programs have successfully controlled diseases like polio and measles, and ongoing efforts aim to control emerging diseases.
- **Hygiene and Sanitation**: Hygiene and sanitation practices, such as regular handwashing, clean water supplies, and waste management, are crucial in preventing disease transmission. Public health campaigns promote these practices to reduce the spread of infectious diseases, especially in areas with limited access to healthcare.
- **Surveillance and Outbreak Response**: Disease surveillance involves monitoring and collecting data on disease incidence and distribution. This information helps detect outbreaks early, allowing public health officials to respond quickly. Outbreak response includes measures such as quarantine, contact tracing, and isolation to contain the spread of infectious agents.

7.6 CHALLENGES IN MICROBIAL PATHOGENICITY AND EPIDEMIOLOGY

Despite advances in understanding microbial pathogenicity and epidemiology, challenges remain in controlling infectious diseases. Drug resistance, emerging pathogens, and changes in environmental and societal factors complicate disease control efforts.

- **Antimicrobial Resistance**: The overuse of antibiotics and other antimicrobials has led to the emergence of drug-resistant pathogens, making infections harder to treat. Resistant infections, such as those caused by *Methicillin-resistant Staphylococcus aureus* (MRSA) or drug-resistant *Mycobacterium tuberculosis*, are a significant public health concern.
- **Emerging and Re-emerging Diseases**: New infectious diseases, such as COVID-19, Ebola, and Zika, continue to emerge, often due to factors like increased human-animal interaction and globalization. Re-emerging diseases, like measles, resurface when vaccination rates decline, highlighting the need for vigilance in disease control.
- **Globalization and Travel**: The rapid movement of people and goods across borders facilitates the spread of infectious diseases. Globalization increases the risk of international outbreaks, as seen with COVID-19, emphasizing the importance of international cooperation in disease surveillance and response.

7.7 CONCLUSION

The study of microbial pathogenicity and epidemiology is crucial for understanding how infectious diseases spread and for developing strategies to control and prevent them. Pathogenicity involves complex interactions between pathogens and hosts, while epidemiology focuses on disease patterns and control measures.

Together, these fields help protect public health by providing the tools and knowledge needed to manage both well-known and emerging infectious diseases.

Addressing the challenges posed by antimicrobial resistance, new pathogens, and global interconnectedness will require continued research, public health interventions, and global cooperation.

12. Conclusion

The study of microbial pathogenicity and epidemiology is crucial for understanding how infectious diseases spread and for developing strategies to control and prevent them. Pathogenicity involves complex interactions between pathogens and hosts, while epidemiology focuses on disease patterns and control measures.

Together, these fields help protect public health by providing the tools and knowledge needed to manage both well-known and emerging infectious diseases.

Addressing the challenges posed by antimicrobial resistance, new pathogens, and global interconnectedness will require continued research, public health interventions, and global cooperation.

CHAPTER 8

MICROBIAL BIOFILMS: CONSEQUENCES FOR HEALTH

Microbial biofilms are complex communities of microorganisms that attach to surfaces and encase themselves in a protective extracellular matrix. These biofilms are ubiquitous, found on a variety of surfaces, including natural environments, industrial settings, and human tissues. While biofilms can play beneficial roles in nature, such as breaking down organic matter in ecosystems, their presence on medical devices and human tissues presents serious health challenges. Biofilms contribute to a range of chronic infections and are notoriously difficult to treat due to their enhanced resistance to antimicrobial agents.

8.1 UNDERSTANDING MICROBIAL BIOFILMS

A biofilm is a syntrophic community of microorganisms in which cells stick to each other and often also to a surface. These adherent cells become embedded within a slimy extracellular matrix that is composed of extracellular polymeric substances (EPSs). The cells within the biofilm produce the EPS components, which are typically a polymeric combination of extracellular polysaccharides, proteins, lipids and DNA. Because they have a three-dimensional structure and represent a community lifestyle for microorganisms, they have been metaphorically described as "cities for microbes".

Biofilms may form on living (biotic) or non-living (abiotic) surfaces and can be common in natural, industrial, and hospital settings. They may constitute a microbiome or be a portion of it. The microbial cells growing in a biofilm are physiologically distinct from planktonic cells of the same organism, which, by contrast, are single cells that may float or swim in a liquid medium. Biofilms can form on the teeth of most animals as dental plaque, where they may cause tooth decay and gum disease.

Microbes form a biofilm in response to a number of different factors, which may include cellular recognition of specific or non-specific attachment sites on a surface, nutritional cues, or in some cases, by exposure of planktonic cells to sub-inhibitory concentrations of antibiotics. A cell that switches to the biofilm mode of growth undergoes a phenotypic shift in behavior in which large suites of genes are differentially regulated.

A biofilm may also be considered a hydrogel, which is a complex polymer that contains many times its dry weight in water. Biofilms are not just bacterial slime layers but biological systems; the bacteria organize themselves into a coordinated functional community. Biofilms can attach to a surface such as a tooth or rock, and may include a single species or a diverse group of microorganisms. Subpopulations of cells within the biofilm differentiate to perform various activities for motility, matrix production, and sporulation, supporting the overall success of the biofilm. The biofilm bacteria can share nutrients and are sheltered from harmful factors in the environment, such as desiccation, antibiotics, and a host body's immune system. A biofilm usually begins to form when a free-swimming, planktonic bacterium attaches to a surface.

8.1.1 Origin of Biofilms

Biofilms are thought to have arisen during primitive Earth as a defense mechanism for prokaryotes, as the conditions at that time were too harsh for their survival. They can be found very early in Earth's fossil records (about 3.25 billion years ago) as both Archaea and Bacteria, and commonly protect prokaryotic cells by providing them with homeostasis, encouraging the development of complex interactions between the cells in the biofilm.

8.1.2 Formation of Biofilms

The formation of a biofilm begins with the attachment of free-floating microorganisms to a surface. The first colonist bacteria of a biofilm may adhere to the surface initially by the weak van der Waals forces and hydrophobic effects. If the colonists are not immediately separated from the surface, they can anchor themselves more permanently using cell adhesion structures such as pili. A unique group of Archaea that inhabit anoxic groundwater have similar structures called hami. Each hamus is a long tube with three hook attachments that are used to attach to each other or to a surface, enabling a community to develop. Hyperthermophilic archaeon *Pyrobaculum calidifontis* produce bundling pili which are homologous to the bacterial TasA filaments, a major component of the extracellular matrix in bacterial biofilms, which contribute to biofilm stability. TasA homologs are encoded by many other archaea, suggesting mechanistic similarities and evolutionary connection between bacterial and archaeal biofilms.

Hydrophobicity can also affect the ability of bacteria to form biofilms. Bacteria with increased hydrophobicity have reduced repulsion between the substratum and the bacterium. Some bacteria species are not able to attach to a surface on their own successfully due to their limited motility but are instead able to anchor themselves to the matrix or directly to other, earlier bacteria colonists. Non-motile bacteria cannot recognize surfaces or aggregate together as easily as motile bacteria.

During surface colonization bacteria cells are able to communicate using quorum sensing (QS) products such as N-acyl homoserine lactone (AHL). Once colonization has begun, the biofilm grows by a combination of cell division and recruitment. Polysaccharide matrices typically enclose bacterial biofilms. The matrix exopolysaccharides can trap QS autoinducers within the biofilm to prevent predator detection and ensure bacterial survival. In addition to the polysaccharides, these matrices may also contain material from the surrounding environment, including but not limited to minerals, soil particles, and blood components, such as erythrocytes and fibrin. The final stage of biofilm formation is known as development, and is the stage in which the biofilm is established and may only change in shape and size.

The development of a biofilm may allow for an aggregate cell colony (or colonies) to be increasingly tolerant or resistant to antibiotics. Cell-cell communication or quorum sensing has been shown to be involved in the formation of biofilm in several bacterial species.

8.1.3 Development

Biofilms are the product of a microbial developmental process. The process is summarized by five major stages of biofilm development:

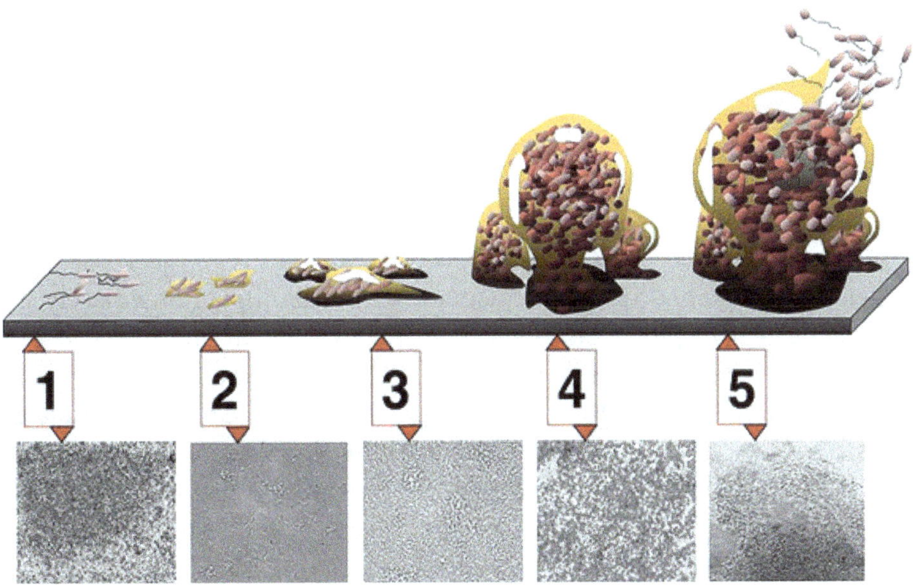

Fig 1: Five Stages of Biofilm Development: 1, Initial Attachment, 2 Irreversible Attachment, 3, Maturation I, 4, Maturation II, and 5< Dispersion.

8.1.4 Dispersal

Dispersal of cells from the biofilm colony is an essential stage of the biofilm life cycle. Dispersal enables biofilms to spread and colonize new surfaces. Enzymes that degrade the biofilm extracellular matrix, such as dispersin B and deoxyribonuclease, may contribute to biofilm dispersal. Enzymes that degrade the biofilm matrix may be useful as anti-biofilm agents. Evidence has shown that a fatty acid messenger, *cis*-2-decenoic acid, is capable of inducing dispersion and inhibiting growth of biofilm colonies. Secreted by *Pseudomonas aeruginosa*, this compound induces cyclo heteromorphic cells in several species of bacteria and the yeast *Candida albicans*. Nitric oxide has also been shown to trigger the dispersal of biofilms of several bacteria species at sub-toxic concentrations. Nitric oxide has potential as a treatment for patients that have chronic infections caused by biofilms.

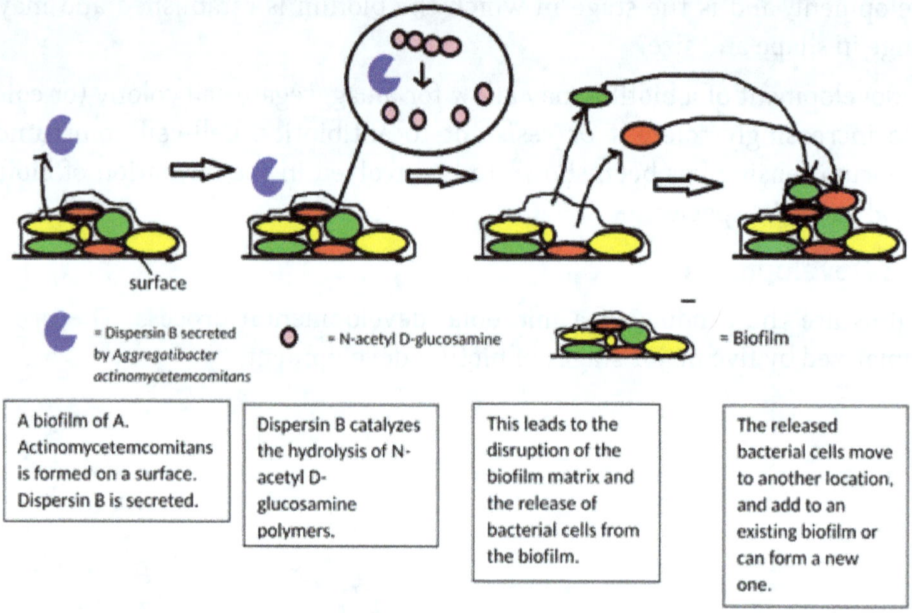

Fig 2: Biofilm Dispersal

It was generally assumed that cells dispersed from biofilms immediately go into the planktonic growth phase. However, studies have shown that the physiology of dispersed cells from *Pseudomonas aeruginosa* biofilms is highly different from that of planktonic and biofilm cells. Hence, the dispersal process is a unique stage during the transition from biofilm to planktonic lifestyle in bacteria. Dispersed cells are found to be highly virulent against macrophages and *Caenorhabditis elegans*, but highly sensitive towards iron stress, as compared with planktonic cells.

Furthermore, *Pseudomonas aeruginosa* biofilms undergo distinct spatiotemporal dynamics during biofilm dispersal or disassembly, with contrasting consequences in recolonization and disease dissemination. Biofilm dispersal induced bacteria to activate dispersal genes to actively depart from biofilms as single cells at consistent velocities but could not recolonize fresh surfaces. In contrast, biofilm disassembly by degradation of a biofilm exopolysaccharide released immotile aggregates at high initial velocities, enabling the bacteria to recolonize fresh surfaces and cause infections in the hosts efficiently.

Hence, biofilm dispersal is more complex than previously thought, where bacterial populations adopting distinct behavior after biofilm departure may be the key to survival of bacterial species and dissemination of diseases.

8.1.5 Properties

Biofilms are usually found on solid substrates submerged in or exposed to an aqueous solution, although they can form as floating mats on liquid surfaces and also on the surface of leaves, particularly in high humidity climates. Given sufficient resources for growth, a biofilm will quickly grow to be macroscopic (visible to the naked eye). Biofilms can contain many different types of microorganism, e.g. bacteria, archaea, protozoa, fungi and algae; each group performs specialized metabolic functions. However, some organisms will form single-species films under certain conditions. The social structure (cooperation/competition) within a biofilm depends highly on the different species present.

Extracellular Matrix

The EPS matrix consists of exopolysaccharides, proteins and nucleic acids. A large proportion of the EPS is more or less strongly hydrated, however, hydrophobic EPS also occur; one example is cellulose which is produced by a range of microorganisms. This matrix encases the cells within it and facilitates communication among them through biochemical signals as well as gene exchange. The EPS matrix also traps extracellular enzymes and keeps them in close proximity to the cells. Thus, the matrix represents an external digestion system and allows for stable synergistic microconsortia of different species. Some biofilms have been found to contain water channels that help distribute nutrients and signalling molecules. This matrix is strong enough that under certain conditions, biofilms can become fossilized (stromatolites).

Bacteria living in a biofilm usually have significantly different properties from free-floating bacteria of the same species, as the dense and protected environment of the film allows them to cooperate and interact in various ways. One benefit of this environment is increased resistance to detergents and antibiotics, as the dense extracellular matrix and the outer layer of cells protect the interior of the community.

In some cases antibiotic resistance can be increased up to 5,000 times. Lateral gene transfer is often facilitated within bacterial and archaeal biofilms and can leads to a more stable biofilm structure. Extracellular DNA is a major structural component of many different microbial biofilms. Enzymatic degradation of extracellular DNA can weaken the biofilm structure and release microbial cells from the surface.

However, biofilms are not always less susceptible to antibiotics. For instance, the biofilm form of *Pseudomonas aeruginosa* has no greater resistance to antimicrobials than do stationary-phase planktonic cells, although when the biofilm is compared to logarithmic-phase planktonic cells, the biofilm does have greater resistance to antimicrobials. This resistance to antibiotics in both stationary-phase cells and biofilms may be due to the presence of persister cells.

8.1.6 Habitats

Biofilms are ubiquitous in organic life. Nearly every species of microorganism have mechanisms by which they can adhere to surfaces and to each other. Biofilms will form on virtually every non-shedding surface in non-sterile aqueous or humid environments. Biofilms can grow in the most extreme environments: from, for example, the extremely hot, briny waters of hot springs ranging from very acidic to very alkaline, to frozen glaciers.

Biofilms can be found on rocks and pebbles at the bottoms of most streams or rivers and often form on the surfaces of stagnant pools of water. Biofilms are important components of food chains in rivers and streams and are grazed by the aquatic invertebrates upon which many fish feed. Biofilms are found on the surface of and inside plants. They can either contribute to crop disease or, as in the case of nitrogen-fixing rhizobia on root nodules, exist symbiotically with the plant. Examples of crop diseases related to biofilms include citrus canker, Pierce's disease of grapes, and bacterial spot of plants such as peppers and tomatoes.

Percolating Filters

Percolating filters in sewage treatment works are highly effective removers of pollutants from settled sewage liquor. They work by trickling the liquid over a bed of hard material which is designed to have a very large surface area. A complex biofilm develops on the surface of the medium which absorbs, adsorbs and metabolises the pollutants. The biofilm grows rapidly and when it becomes too thick to retain its grip on the media it washes off and is replaced by newly grown film. The washed off ("sloughed" off) film is settled out of the liquid stream to leave a highly purified effluent.

Slow Sand Filter

Slow sand filters are used in water purification for treating raw water to produce a potable product. They work through the formation of a biofilm called the hypogeal layer or *Schmutzdecke* in the top few millimetres of the fine sand layer. The *Schmutzdecke* is formed in the first 10–20 days of operation and consists of bacteria, fungi, protozoa, rotifera and a range of aquatic insect larvae. As an epigeal biofilm ages, more algae tend to develop and larger aquatic organisms may be present including some bryozoa, snails and annelid worms. The surface biofilm is the layer that provides the effective purification in potable water treatment, the underlying sand providing the support medium for this biological treatment layer. As water passes through the hypogeal layer, particles of foreign matter are trapped in the mucilaginous matrix and soluble organic material is adsorbed. The contaminants are metabolised by the bacteria, fungi and protozoa. The water produced from an exemplary slow sand filter is of excellent quality with 90–99% bacterial cell count reduction.

Rhizosphere

Plant-beneficial microbes can be categorized as plant growth-promoting rhizobacteria. These plant growth-promoters colonize the roots of plants, and provide a wide range of beneficial functions for their host including nitrogen fixation, pathogen suppression, anti-fungal properties, and the breakdown of organic materials. One of these functions is the defense against pathogenic, soil-borne bacteria and fungi by way of induced systemic resistance (ISR) or induced systemic responses triggered by pathogenic microbes (pathogen-induced systemic acquired resistance). Plant exudates act as chemical signals for host specific bacteria to colonize. Rhizobacteria colonization steps include attractions, recognition, adherence, colonization, and growth. Bacteria that have been shown to be beneficial and form biofilms include *Bacillus, Pseudomonas,* and *Azospirillum.* Biofilms in the rhizosphere often result in pathogen or plant induced systemic resistances. Molecular properties on the surface of the bacterium cause an immune response in the plant host. These microbe associated molecules interact with receptors on the surface of plant cells, and activate a biochemical response that is thought to include several different genes at a number of loci. Several other signaling molecules have been linked to both induced systemic responses and pathogen-induced systemic responses, such as jasmonic acid and ethylene. Cell envelope components such as bacterial flagella and lipopolysaccharides, which are recognized by plant cells as components of pathogens. Certain iron metabolites produced by *Pseudomonas* have also been shown to create an induced systemic response. This function of the biofilm helps plants build stronger resistance to pathogens.

Plants that have been colonized by PGPR forming a biofilm have gained systemic resistances and are primed for defense against pathogens. This means that the genes necessary for the production of proteins that work towards defending the plant against pathogens have been expressed, and the plant has a "stockpile" of compounds to release to fight off pathogens. A primed defense system is much faster in responding to pathogen induced infection, and may be able to deflect pathogens before they are able to establish themselves. Plants increase the production of lignin, reinforcing cell walls and making it difficult for pathogens to penetrate into the cell, while also cutting off nutrients to already infected cells, effectively halting the invasion. They produce antimicrobial compounds such as phytoalexins, chitinases, and proteinase inhibitors, which prevent the growth of pathogens. These functions of disease suppression and pathogen resistance ultimately lead to an increase in agricultural production and a decrease in the use of chemical pesticides, herbicides, and fungicides because there is a reduced amount of crop loss due to disease. Induced systemic resistance and pathogen-induced systemic acquired resistance are both potential functions of biofilms in the rhizosphere, and should be taken into consideration when applied to new age agricultural practices because of their effect on disease suppression without the use of dangerous chemicals.

Mammalian Gut

Studies in 2003 discovered that the immune system supports biofilm development in the large intestine. This was supported mainly with the fact that the two most abundantly produced molecules by the immune system also support biofilm production and are associated with the biofilms developed in the gut. This is especially important because the appendix holds a mass amount of these bacterial biofilms. This discovery helps to distinguish the possible function of the appendix and the idea that the appendix can help reinoculate the gut with good gut flora. However, modified or disrupted states of biofilms in the gut have been connected to diseases such as inflammatory bowel disease and colorectal cancer.

Human Environment

In the human environment, biofilms can grow in showers very easily since they provide a moist and warm environment for them to thrive. They can form inside water and sewage pipes and cause clogging and corrosion. On floors and counters, they can make sanitation difficult in food preparation areas. In soil, they can cause bioclogging. In cooling- or heating-water systems, they are known to reduce heat transfer. Biofilms in marine engineering systems, such as pipelines of the offshore oil and gas industry, can lead to substantial corrosion problems. Corrosion is mainly due to abiotic factors; however, at least 20% of corrosion is caused by microorganisms that are attached to the metal subsurface (i.e., microbially influenced corrosion).

Ship Fouling

Bacterial adhesion to boat hulls serves as the foundation for biofouling of seagoing vessels. Once a film of bacteria forms, it is easier for other marine organisms such as barnacles to attach. Such fouling can reduce maximum vessel speed by up to 20%, prolonging voyages and consuming fuel. Time in dry dock for refitting and repainting reduces the productivity of shipping assets, and the useful life of ships is also reduced due to corrosion and mechanical removal (scraping) of marine organisms from ships' hulls.

Stromatolites

Stromatolites are layered accretionary structures formed in shallow water by the trapping, binding and cementation of sedimentary grains by microbial biofilms, especially of cyanobacteria. Stromatolites include some of the most ancient records of life on Earth, and are still forming today.

Dental Plaque

Within the human body, biofilms are present on the teeth as dental plaque, where they may cause tooth decay and gum disease. These biofilms can either be in an uncalcified state that can be removed by dental instruments, or a calcified state which is more difficult to remove. Removal techniques can also include antimicrobials.

Dental plaque is an oral biofilm that adheres to the teeth and consists of many species of both bacteria and fungi (such as *Streptococcus mutans* and *Candida albicans*), embedded in salivary polymers and microbial extracellular products. The accumulation of microorganisms subjects the teeth and gingival tissues to high concentrations of bacterial metabolites which results in dental disease.

Biofilm on the surface of teeth is frequently subject to oxidative stress and acid stress. Dietary carbohydrates can cause a dramatic decrease in pH in oral biofilms to values of 4 and below (acid stress). A pH of 4 at body temperature of 37 °C causes depurination of DNA, leaving apurinic (AP) sites in DNA, especially loss of guanine.

Dental plaque biofilm can result in dental caries if it is allowed to develop over time. An ecologic shift away from balanced populations within the dental biofilm is driven by certain (cariogenic) microbiological populations beginning to dominate when the environment favors them.

The shift to an acidogenic, aciduric, and cariogenic microbiological population develops and is maintained by frequent consumption of fermentable dietary carbohydrate. The resulting activity shift in the biofilm (and resulting acid production within the biofilm, at the tooth surface) is associated with an imbalance of demineralization over remineralization, leading to net mineral loss within dental hard tissues (enamel and then dentin), the symptom being a carious lesion, or cavity. By preventing the dental plaque biofilm from maturing or by returning it back to a non-cariogenic state, dental caries can be prevented and arrested. This can be achieved through the behavioral step of reducing the supply of fermentable carbohydrates (i.e., sugar intake) and frequent removal of the biofilm (i.e., toothbrushing).

Intercellular Communication

A peptide pheromone quorum sensing signaling system in *S. mutans* includes the competence stimulating peptide (CSP) that controls genetic competence. Genetic competence is the ability of a cell to take up DNA released by another cell. Competence can lead to genetic transformation, a form of sexual interaction, favored under conditions of high cell density and/or stress where there is maximal opportunity for interaction between the competent cell and the DNA released from nearby donor cells. This system is optimally expressed when *S. mutans* cells reside in an actively growing biofilm. Biofilm grown *S. mutans* cells are genetically transformed at a rate 10- to 600-fold higher than *S. mutans* growing as free-floating planktonic cells suspended in liquid.

When the biofilm, containing *S. mutans* and related oral streptococci, is subjected to acid stress, the competence regulon is induced, leading to resistance to being killed by acid. As pointed out by Michod et al., transformation in bacterial pathogens likely provides for effective and efficient recombinational repair of DNA damages. It appears that *S. mutans* can survive the frequent acid stress in oral biofilms, in part, through the recombinational repair provided by competence and transformation.

8.1.7 Predator-Prey Interactions

Predator-prey interactions between biofilms and bacterivores, such as the soil-dwelling nematode *Caenorhabditis elegans,* had been extensively studied. Via the production of sticky matrix and formation of aggregates, *Yersinia pestis* biofilms can prevent feeding by obstructing the mouth of *C. elegans*. Moreover, *Pseudomonas aeruginosa* biofilms can impede the slithering motility of *C. elegans*, termed as 'quagmire phenotype', resulting in trapping of *C. elegans* within the biofilms and preventing the exploration of nematodes to feed on susceptible biofilms. This significantly reduced the ability of predator to feed and reproduce, thereby promoting the survival of biofilms. *Pseudomonas aeruginosa* biofilms can also mask their chemical signatures, where they reduced the diffusion of quorum sensing molecules into the environment and prevented the detection of *C. elegans*.

8.1.8 Taxonomic Diversity

Many different bacteria form biofilms, including gram-positive (e.g. *Bacillus* spp, *Listeria monocytogenes*, *Staphylococcus* spp, and lactic acid bacteria, including *Lactobacillus plantarum* and *Lactococcus lactis*) and gram-negative species (e.g. *Escherichia coli*, or *Pseudomonas aeruginosa*). Cyanobacteria also form biofilms in aquatic environments.

Biofilms are formed by bacteria that colonize plants, e.g. *Pseudomonas putida*, *Pseudomonas fluorescens*, and related pseudomonads which are common plant-associated bacteria found on leaves, roots, and in the soil, and the majority of their natural isolates form biofilms. Several nitrogen-fixing symbionts of legumes such as *Rhizobium leguminosarum* and *Sinorhizobium meliloti* form biofilms on legume roots and other inert surfaces.

Along with bacteria, biofilms are also generated by archaea and by a range of eukaryotic organisms, including fungi e.g. *Cryptococcus laurentii* and microalgae. Among microalgae, one of the main progenitors of biofilms are diatoms, which colonise both fresh and marine environments worldwide.

8.1.9 Infectious Diseases

Biofilms have been found to be involved in a wide variety of microbial infections in the body, by one estimate 80% of all infections. Infectious processes in which biofilms have been implicated include common problems such as bacterial vaginosis, urinary tract infections, catheter infections, middle-ear infections, formation of dental plaque, gingivitis, coating contact lenses, and less common but more lethal processes such as endocarditis, infections in cystic fibrosis, and infections of permanent indwelling devices such as joint prostheses, heart valves, and intervertebral disc. The first visual evidence of a biofilm was recorded after spine surgery. It was found that in the absence of clinical presentation of infection, impregnated bacteria could form a biofilm around an implant, and this biofilm can remain undetected via contemporary diagnostic methods, including swabbing. Implant biofilm is frequently present in "aseptic" pseudarthrosis cases. Furthermore, it has been noted that bacterial biofilms may impair cutaneous wound healing and reduce topical antibacterial efficiency in healing or treating infected skin wounds. The diversity of *P. aeruginosa* cells within a biofilm is thought to make it harder to treat the infected lungs of people with cystic fibrosis. Early detection of biofilms in wounds is crucial to successful chronic wound management. Although many techniques have developed to identify planktonic bacteria in viable wounds, few have been able to quickly and accurately identify bacterial biofilms. Future studies are needed to find means of identifying and monitoring biofilm colonization at the bedside to permit timely initiation of treatment.

It has been shown that biofilms are present on the removed tissue of 80% of patients undergoing surgery for chronic sinusitis. The patients with biofilms were shown to have been denuded of cilia and goblet cells, unlike the controls without biofilms who had normal cilia and goblet cell morphology. Biofilms were also found on samples from two of 10 healthy controls mentioned. The species of bacteria from intraoperative cultures did not correspond to the bacteria species in the biofilm on the respective patient's tissue. In other words, the cultures were negative though the bacteria were present. New staining techniques are being developed to differentiate bacterial cells growing in living animals, e.g. from tissues with allergy-inflammations.

Research has shown that sub-therapeutic levels of β-lactam antibiotics induce biofilm formation in *Staphylococcus aureus*. This sub-therapeutic level of antibiotic may result from the use of antibiotics as growth promoters in agriculture, or during the normal course of antibiotic therapy. The biofilm formation induced by low-level methicillin was inhibited by DNase, suggesting that the sub-therapeutic levels of antibiotic also induce extracellular DNA release. Moreover, from an evolutionary point of view, the creation of the tragedy of the commons in pathogenic microbes may provide advanced therapeutic ways for chronic infections caused by biofilms via genetically engineered invasive cheaters who can invade wild-types 'cooperators' of pathogenic bacteria until cooperator populations go to extinction or overall population 'cooperators and cheaters ' go to extinction.

Pseudomonas Aeruginosa

P. aeruginosa represents a commonly used biofilm model organism since it is involved in different types of biofilm-associated chronic infections. Examples of such infections include chronic wounds, chronic otitis media, chronic prostatitis and chronic lung infections in cystic fibrosis (CF) patients. About 80% of CF patients have chronic lung infection, caused mainly by *P. aeruginosa* growing in a non-surface attached biofilms surround by PMN. The infection remains present despite aggressive antibiotic therapy and is a common cause of death in CF patients due to constant inflammatory damage to the lungs. In patients with CF, one therapy for treating early biofilm development is to employ DNase to structurally weaken the biofilm.

Biofilm formation of *P. aeruginosa*, along with other bacteria, is found in 90% of chronic wound infections, which leads to poor healing and high cost of treatment estimated at more than US$25 billion every year in the United States. In order to minimize the *P. aeruginosa* infection, host epithelial cells secrete antimicrobial peptides, such as lactoferrin, to prevent the formation of the biofilms.

Streptococcus Pneumoniae

Streptococcus pneumoniae is the main cause of community-acquired pneumonia and meningitis in children and the elderly, and of sepsis in HIV-infected persons. When *S. pneumoniae* grows in biofilms, genes are specifically expressed that respond to oxidative stress and induce competence. Formation of a biofilm depends on competence stimulating peptide (CSP). CSP also functions as a quorum-sensing peptide. It not only induces biofilm formation, but also increases virulence in pneumonia and meningitis.

It has been proposed that competence development and biofilm formation is an adaptation of *S. pneumoniae* to survive the defenses of the host. In particular, the host's polymorphonuclear leukocytes produce an oxidative burst to defend against the invading bacteria, and this response can kill bacteria by damaging their DNA. Competent *S. pneumoniae* in a biofilm have the survival advantage that they can more easily take up transforming DNA from nearby cells in the biofilm to use for recombinational repair of oxidative damages in their DNA. Competent *S. pneumoniae* can also secrete an enzyme (murein hydrolase) that destroys non-competent cells (fratricide) causing DNA to be released into the surrounding medium for potential use by the competent cells.

The insect antimicrobial peptide cecropin A can destroy planktonic and sessile biofilm-forming uropathogenic *E. coli* cells, either alone or when combined with the antibiotic nalidixic acid, synergistically clearing infection in vivo (in the insect host *Galleria mellonella*) without off-target cytotoxicity. The multi-target mechanism of action involves outer membrane permeabilization followed by biofilm disruption triggered by the inhibition of efflux pump activity and interactions with extracellular and intracellular nucleic acids.

Escherichia Coli

Escherichia coli biofilms are responsible for many intestinal infectious diseases. The Extraintestinal group of *E. coli* (ExPEC) is the dominant bacterial group that attacks the urinary system, which leads to urinary tract infections. The biofilm formation of these pathogenic *E. coli* is hard to eradicate due to the complexity of its aggregation structure, and it has a significant contribution to developing aggressive medical complications, increase in hospitalization rate, and cost of treatment. The development of *E. coli* biofilm is a common leading cause of urinary tract infections (UTI) in hospitals through its contribution to developing medical device-associated infections. Catheter-associated urinary tract infections (CAUTI) represent the most common hospital-acquired infection due to the formation of the pathogenic *E. coli biofilm* inside the catheters.

Staphylococcus Aureus

Staphylococcus aureus pathogen can attack skin and lungs, leading to skin infection and pneumonia. Moreover, the biofilm infections network of *S. aureus* plays a critical role in preventing immune cells, such as macrophages from eliminating and destroying bacterial cells. Furthermore, biofilm formation by bacteria, such as *S. aureus*, not only develops resistance against antibiotic medication but also develop internal resistance toward antimicrobial peptides (AMPs), leading to preventing the inhibition of the pathogen and maintaining its survival.

Serratia Marcescens

Serratia marcescens is a fairly common opportunistic pathogen that can form biofilms on various surfaces, including medical devices such as catheters and implants, as well as natural environments like soil and water. The formation of biofilms by *S. marcescens* is a serious concern because of its ability to adhere to and colonize surfaces, protecting itself from host immune responses and antimicrobial agents. This strength makes infections caused by *S. marcescens* challenging to treat, specifically in hospitals where the bacterium can cause severe, and specific, infections.

Research suggests that biofilm formation by S. marcescens is a process controlled by both nutrient cues and the quorum-sensing system. Quorum sensing influences the bacterium's ability to adhere to surfaces and establish mature biofilms, whereas the availability of specific nutrients can enhance or inhibit biofilm development.

S. marcescens creates biofilms that ultimately develop into a highly porous, thread-like structure composed of chains of cells, filaments, and cell clusters. Research has shown that *S. marcescens* biofilms exhibit complex structural organization, including the formation of microcolonies and channels that facilitate nutrient and waste exchange. The production of extracellular polymeric substances (EPS) is a key factor in biofilm development, contributing to the bacterium's adhesion and resistance to antimicrobial agents. In addition to its role in healthcare-associated infections, *S. marcescens* biofilms have been implicated in the deterioration of industrial equipment and processes. For example, biofilm growth in cooling towers can lead to biofouling and reduced efficiency.

Efforts to control and prevent biofilm formation by *S. marcescens* involve the use of antimicrobial coatings on medical devices, the development of targeted biofilm disruptors, and improved sterilization protocols. Further research into the molecular mechanisms governing *S. marcescens* biofilm formation and persistence is crucial for developing effective strategies to combat its associated risks. The use of indole compounds has been studied to be used as protection against biofilm formation.

8.1.10 Uses and Impact

In Medicine

It is suggested that around two-thirds of bacterial infections in humans involve biofilms. Infections associated with the biofilm growth usually are challenging to eradicate. This is mostly due to the fact that mature biofilms display antimicrobial tolerance, and immune response evasions. Biofilms often form on the inert surfaces of implanted devices such as catheters, prosthetic cardiac valves and intrauterine devices. Some of the most difficult infections to treat are those associated with the use of medical devices.

The rapidly expanding worldwide industry for biomedical devices and tissue engineering related products is already at $180 billion per year, yet this industry continues to suffer from microbial colonization. No matter the sophistication, microbial infections can develop on all medical devices and tissue engineering constructs. 60-70% of hospital-acquired infections are associated with the implantation of a biomedical device. This leads to 2 million cases annually in the U.S., costing the healthcare system over $5 billion in additional healthcare expenses.

The level of antibiotic resistance in a biofilm is much greater than that of non-biofilm bacteria, and can be as much as 5,000 times greater. The extracellular matrix of biofilm is considered one of the leading factors that can reduce the penetration of antibiotics into a biofilm structure and contributes to antibiotic resistance. Further, it has been demonstrated that the evolution of resistance to antibiotics may be affected by the biofilm lifestyle. Bacteriophage therapy can disperse the biofilm generated by antibiotic-resistant bacteria.

It has been shown that the introduction of a small current of electricity to the liquid surrounding a biofilm, together with small amounts of antibiotic can reduce the level of antibiotic resistance to levels of non-biofilm bacteria. This is termed the bioelectric effect. The application of a small DC current on its own can cause a biofilm to detach from its surface. A study showed that the type of current used made no difference to the bioelectric effect.

In Industry

Biofilms can also be harnessed for constructive purposes. For example, many sewage treatment plants include a secondary treatment stage in which waste water passes over biofilms grown on filters, which extract and digest organic compounds. In such biofilms, bacteria are mainly responsible for removal of organic matter (BOD), while protozoa and rotifers are mainly responsible for removal of suspended solids (SS), including pathogens and other microorganisms. Slow sand filters rely on biofilm development in the same way to filter surface water from lake, spring or river sources for drinking purposes. What is regarded as clean water is effectively a waste material to these microcellular organisms. Biofilms can help eliminate petroleum oil from contaminated oceans or marine systems. The oil is eliminated by the hydrocarbon-degrading activities of communities of hydrocarbonoclastic bacteria (HCB). Biofilms are used in microbial fuel cells (MFCs) to generate electricity from a variety of starting materials, including complex organic waste and renewable biomass. Biofilms are also relevant for the improvement of metal dissolution in bioleaching industry, and aggregation of microplastics pollutants for convenient removal from the environment.

Food Industry

Biofilms have become problematic in several food industries due to the ability to form on plants and during industrial processes. Bacteria can survive long periods of time in water, animal manure, and soil, causing biofilm formation on plants or in the processing equipment. The buildup of biofilms can affect the heat flow across a surface and increase surface corrosion and frictional resistance of fluids. These can lead to a loss of energy in a system and overall loss of products. Along with economic problems, biofilm formation on food poses a health risk to consumers due to the ability to make the food more resistant to disinfectants As a result, from 1996 to 2010 the Centers for Disease Control and Prevention estimated 48 million foodborne illnesses per year. Biofilms have been connected to about 80% of bacterial infections in the United States.

In produce, microorganisms attach to the surfaces and biofilms develop internally. During the washing process, biofilms resist sanitization and allow bacteria to spread across the produce, especially via kitchen utensils. This problem is also found in ready-to-eat foods, because the foods go through limited cleaning procedures before consumption Due to the perishability of dairy products and limitations in cleaning procedures, resulting in the buildup of bacteria, dairy is susceptible to biofilm formation and contamination. The bacteria can spoil the products more readily and contaminated products pose a health risk to consumers. One species of bacteria that can be found in various industries and is a major cause of foodborne disease is *Salmonella*. Large amounts of Salmonella contamination can be found in the poultry processing industry as about 50% of *Salmonella* strains can produce biofilms on poultry farms. *Salmonella* increases the risk of foodborne illnesses when the poultry products are not cleaned and cooked correctly. *Salmonella* is also found in the seafood industry where biofilms form from seafood borne pathogens on the seafood itself as well as in water. Shrimp products are commonly affected by *Salmonella* because of unhygienic processing and handling techniques The preparation practices of shrimp and other seafood products can allow for bacteria buildup on the products.

New forms of cleaning procedures are being tested to reduce biofilm formation in these processes which will lead to safer and more productive food processing industries. These new forms of cleaning procedures also have a profound effect on the environment, often releasing toxic gases into the groundwater reservoirs. As a response to the aggressive methods employed in controlling biofilm formation, there are a number of novel technologies and chemicals under investigation that can prevent either the proliferation or adhesion of biofilm-secreting microbes. Latest proposed biomolecules presenting marked anti-biofilm activity include a range of metabolites such as bacterial rhamnolipids and even plant- and animal-derived alkaloids.

In Aquaculture

In shellfish and algal aquaculture, biofouling microbial species tend to block nets and cages and ultimately outcompete the farmed species for space and food.

Bacterial biofilms start the colonization process by creating microenvironments that are more favorable for biofouling species. In the marine environment, biofilms could reduce the hydrodynamic efficiency of ships and propellers, lead to pipeline blockage and sensor malfunction, and increase the weight of appliances deployed in seawater. Numerous studies have shown that biofilm can be a reservoir for potentially pathogenic bacteria in freshwater aquaculture. Moreover, biofilms are important in establishing infections on the fish. As mentioned previously, biofilms can be difficult to eliminate even when antibiotics or chemicals are used in high doses. The role that biofilm plays as reservoirs of bacterial fish pathogens has not been explored in detail but it certainly deserves to be studied.

8.1.11 Eukaryotic

Along with bacteria, biofilms are often initiated and produced by eukaryotic microbes. The biofilms produced by eukaryotes is usually occupied by bacteria and other eukaryotes alike, however the surface is cultivated and EPS is secreted initially by the eukaryote. Both fungi and microalgae are known to form biofilms in such a way. Biofilms of fungal origin are important aspects of human infection and fungal pathogenicity, as the fungal infection is more resistant to antifungals.

In the environment, fungal biofilms are an area of ongoing research. One key area of research is fungal biofilms on plants. For example, in the soil, plant associated fungi including mycorrhiza have been shown to decompose organic matter and protect plants from bacterial pathogens.

Biofilms in aquatic environments are often founded by diatoms. The exact purpose of these biofilms is unknown, however there is evidence that the EPS produced by diatoms facilitates both cold and salinity stress. These eukaryotes interact with a diverse range of other organisms within a region known as the phycosphere, but importantly are the bacteria associated with diatoms, as it has been shown that although diatoms excrete EPS, they only do so when interacting with certain bacteria species.

8.1.12 Horizontal Gene Transfer

Horizontal gene transfer is the lateral transfer of genetic material between cellular organisms. It happens frequently in prokaryotes, and less frequently in eukaryotes. In bacteria, horizontal gene transfer can occur through transformation (uptake of free floating DNA in the environment), transduction (virus mediated DNA uptake), or conjugation (transfer of DNA between pili structures of two adjacent bacteria). Recent studies have also uncovered other mechanisms, such as membrane vesicle transmission or gene transfer agents. Biofilms promote horizontal gene transfer in a variety of ways.

Bacterial conjugation has been shown to accelerate biofilm formation in difficult environment due to the robust connections established by the conjugative pili. These connections can often foster cross-species transfer events due to the diverse heterogeneity of many biofilms. Additionally, biofilms are structurally confined by a polysaccharide matrix, providing the close spatial requirements for conjugation. Transformation is also frequently observed in biofilms. Bacterial autolysis is a key mechanism in biofilm structural regulation, providing an abundant source of competent DNA primed for transformative uptake. In some instances, inter-biofilm quorum sensing can enhance the competence of free floating eDNA, further promoting transformation.

Stx gene transfer through bacteriophage carriers has been witnessed within biofilms, which suggests that biofilms are also a suitable environment for transduction. Membrane vesicles HGT occurs when released membrane vesicles (containing genetic information) fuse with a recipient bacteria, and release genetic material into the bacteria's cytoplasm. Recent research has revealed that membrane vesicle HGT can promote single-strain biofilm formation, yet the role membrane vesicle HGT plays in the formation of multistrain biofilms is still unknown. GTAs, or gene transfer agents, are phage-like particles produced by the host bacteria and contain random DNA fragments from the host bacteria genome. HGT within biofilms can confer antibiotic resistance or increased pathogenicity across the biofilms' population, promoting biofilm homeostasis.

Examples

Conjugative plasmids may encode biofilm-associated proteins, such as PtgA, PrgB, or PrgC which promote cell adhesion (required for early biofilm formation). Genes encoding type III fimbriae are found in pOLA52 (*Klebsiella pneumoniae* plasmid) which promote conjugative-pilus-dependent biofilm formation.

Transformation commonly occurs within biofilms. A phenomenon called fratricide can be seen among streptococcal species in which cell-wall degrading enzymes are released, lysing neighboring bacteria and releasing their DNA. This DNA can then be taken up by the surviving bacteria (transformation). Competence stimulating peptides may play an important role in biofilm formation among *S. pneumoniae* and *S. mutans* as well. Among *V. cholerae*, the competence pilus itself promotes cell aggregation through pilus-pilus interactions at the beginning of biofilm formation.

Phage invasion may play a role in biofilm life cycles, lysing bacteria and releasing their eDNA, which strengthens biofilm structures and can be taken up by neighboring bacteria in transformation. Biofilm destruction caused by the *E. coli* phage Rac and the *P. aeruginosa* prophage Pf4 causes detachment of cells from the biofilm. Detachment is a biofilm phenomenon which requires more study, but is hypothesized to proliferate the bacterial species that comprise the biofilm.

Membrane vesicle HGT has been witnessed occurring in marine environments, among *Neisseria gonorrhoeae, Pseudomonas aeruginosa, Helicobacter pylori,* and among many other bacterial species. Even though membrane vesicle HGT has been shown as a contributing factor in biofilm formation, research is still required to prove that membrane vesicle mediated HGT occurs within biofilms. Membrane vesicle HGT has also been shown to modulate phage-bacteria interactions in *Bacillus subtilis* SPP1 phage-resistant cells (lacking the SPP1 receptor protein). Upon exposure to vesicles containing receptors, transduction of pBT163 (a cat-encoding plasmid) occurs, resulting in the expression of the SPP1 receptor protein, opening the receptive bacteria to future phage infection.

Recent research has shown that the archaeal species *H. volcanii* has some biofilm phenotypes similar to bacterial biofilms such as differentiation and HGT, which required cell-cell contact and involved formation of cytosolic bridges and cellular fusion events.

8.2 BIOFILMS AND CHRONIC INFECTIONS

One of the most concerning aspects of microbial biofilms is their ability to cause chronic infections. Unlike acute infections, which are generally caused by planktonic (free-floating) bacteria, biofilm-associated infections are persistent and difficult to eradicate.

- **Biofilms on Medical Devices:** Biofilms frequently form on medical devices, including catheters, heart valves, dental implants, and orthopedic implants. When biofilms form on these surfaces, they can lead to persistent infections that are resistant to treatment, often requiring the removal of the infected device. For instance, urinary tract infections associated with catheter use are commonly due to biofilm formation by bacteria like *Escherichia coli* and *Pseudomonas aeruginosa.*
- **Respiratory Infections:** In patients with cystic fibrosis or chronic obstructive pulmonary disease (COPD), biofilms form in the respiratory tract, causing persistent and recurrent infections. *Pseudomonas aeruginosa*, a common pathogen in cystic fibrosis patients, can establish robust biofilms in the lungs, leading to inflammation, lung damage, and decreased respiratory function.
- **Chronic Wounds and Skin Infections:** Biofilms also play a role in chronic wound infections, such as diabetic foot ulcers and bedsores. These wounds are difficult to heal because the biofilm matrix protects bacteria from the immune response and antibiotics. *Staphylococcus aureus* and *Streptococcus pyogenes* are among the common biofilm-forming pathogens in chronic wounds.

8.3 ENHANCED RESISTANCE TO ANTIBIOTICS AND HOST DEFENSES

Biofilm-associated infections are notoriously difficult to treat due to the enhanced resistance of biofilm communities to antibiotics and immune responses. This resistance is a major factor in the persistence and recurrence of biofilm-related infections.

- **Antibiotic Resistance Mechanisms:** Microbes in biofilms exhibit higher resistance to antibiotics compared to their planktonic counterparts. The EPS matrix acts as a physical barrier, limiting the penetration of antibiotics. Additionally, within the biofilm, bacteria can enter a slow-growing or dormant state that makes them less susceptible to antibiotics, which often target actively dividing cells. Some bacteria within biofilms also possess specific resistance genes, further complicating treatment efforts.

- **Immune System Evasion**: Biofilms protect pathogens from the host immune system. The biofilm structure hinders immune cells, such as neutrophils, from effectively reaching and killing bacterial cells. Moreover, bacteria within biofilms can avoid being recognized by the immune system due to alterations in surface molecules. This evasion results in chronic, low-level infections that cause ongoing inflammation and tissue damage.

8.4 CONSEQUENCES FOR PUBLIC HEALTH

The presence of biofilms in healthcare settings has significant implications for public health. Biofilm-related infections are associated with increased healthcare costs, longer hospital stays, and higher morbidity and mortality rates. Addressing biofilms is a priority in infection control and healthcare-associated infection (HAI) prevention.

- **Healthcare-Associated Infections (HAIs)**: Biofilms are a major contributor to HAIs, which affect millions of patients globally each year. HAIs associated with biofilms are particularly concerning because of their persistence and resistance to standard treatments, often necessitating more aggressive interventions, such as surgical removal of infected devices.

- **Economic Burden**: Biofilm-associated infections increase healthcare costs due to extended treatments, repeated hospitalizations, and additional surgeries. Chronic infections require more resources for management and rehabilitation, placing a burden on healthcare systems and patients alike.

- **Patient Impact:** Patients with biofilm-associated infections often experience prolonged pain, discomfort, and a reduced quality of life. Chronic infections can lead to complications, long-term health issues, and in severe cases, mortality.

8.5 APPROACHES TO CONTROL AND PREVENT BIOFILM-RELATED INFECTIONS

The challenge of treating biofilm-associated infections has led researchers and healthcare providers to explore various strategies for prevention, control, and eradication of biofilms. These strategies include both traditional antimicrobial approaches and innovative therapies.

- **Preventing Biofilm Formation on Medical Devices:** Preventative measures include using materials that resist biofilm formation for medical devices, such as coatings with antimicrobial properties. Silver and other metal-based coatings have shown promise in reducing biofilm formation on devices. Regular cleaning and sterilization protocols are also essential to prevent biofilm buildup on medical equipment.
- **Antibiotic Combination Therapy:** Since single antibiotics are often ineffective against biofilms, combination therapy using multiple antibiotics or antibiotics with different modes of action may improve treatment outcomes. However, the choice of antibiotics and their dosages must be carefully managed to prevent the development of further resistance.
- **Innovative Therapies:** Novel approaches for combating biofilms include the use of enzymes that degrade the EPS matrix, making biofilms more susceptible to antibiotics. Bacteriophages, which are viruses that infect and kill bacteria, are also being explored as a potential therapy for biofilm-related infections. Additionally, researchers are investigating the use of nanoparticles and other nanomaterials as effective agents for disrupting biofilms.

8.6 FUTURE DIRECTIONS AND CHALLENGES IN BIOFILM RESEARCH

The persistence of biofilm-related infections highlights the need for ongoing research and innovation in this field. Despite advancements, several challenges remain in fully understanding and effectively treating biofilm-associated infections.

- **Improving Detection Techniques:** Detecting biofilms in clinical settings remains a challenge, as standard laboratory methods may not reveal the presence of biofilms. Advances in imaging and molecular diagnostics are crucial for accurately identifying biofilm-associated infections and assessing their extent.

- **Understanding Biofilm Dynamics:** Biofilms are dynamic environments with complex interactions among microbial communities. Future research should focus on understanding the molecular mechanisms that regulate biofilm formation, maintenance, and dispersion. This knowledge will be essential for developing targeted therapies that can disrupt biofilms at different stages of their development.
- **Addressing Antimicrobial Resistance:** The rise of antimicrobial resistance (AMR) in biofilm-associated infections is a pressing issue. Researchers must continue to develop new antimicrobial agents and alternative treatments that can effectively target biofilms while minimizing the risk of resistance.

8.7 CONCLUSION

Microbial biofilms present a formidable challenge in healthcare due to their resilience, antibiotic resistance, and ability to evade immune defenses. Biofilm-associated infections are particularly problematic on medical devices, in chronic wounds, and in respiratory and urinary tracts, leading to significant health risks, economic burdens, and impacts on patient quality of life. Addressing these issues requires a multifaceted approach, from improving prevention strategies to advancing therapeutic interventions. Future research and innovative treatments are essential for overcoming the challenges posed by biofilms and reducing their impact on public health. As biofilm research progresses, new insights and technologies will enhance our ability to manage and control these complex microbial communities, ultimately improving patient outcomes and advancing infection control practices.

CHAPTER 9
IMMUNOLOGY

Immunology is the branch of biomedical science that focuses on the study of the immune system, its structure, functions, and disorders. The immune system plays a crucial role in protecting the body from harmful pathogens such as bacteria, viruses, and fungi, while also monitoring and eliminating damaged or abnormal cells. A robust immune response is essential for maintaining health, and any disruption in this system can lead to diseases ranging from infections to autoimmune disorders and even cancer.

9.1 THE IMMUNE SYSTEM: AN OVERVIEW

The immune system is a network of biological systems that protects an organism from diseases. It detects and responds to a wide variety of pathogens, from viruses to parasitic worms, as well as cancer cells and objects such as wood splinters, distinguishing them from the organism's own healthy tissue. Many species have two major subsystems of the immune system. The innate immune system provides a preconfigured response to broad groups of situations and stimuli. The adaptive immune system provides a tailored response to each stimulus by learning to recognize molecules it has previously encountered. Both use molecules and cells to perform their functions.

Nearly all organisms have some kind of immune system. Bacteria have a rudimentary immune system in the form of enzymes that protect against viral infections. Other basic immune mechanisms evolved in ancient plants and animals and remain in their modern descendants. These mechanisms include phagocytosis, antimicrobial peptides called defensins, and the complement system. Jawed vertebrates, including humans, have even more sophisticated defense mechanisms, including the ability to adapt to recognize pathogens more efficiently. Adaptive (or acquired) immunity creates an immunological memory leading to an enhanced response to subsequent encounters with that same pathogen. This process of acquired immunity is the basis of vaccination.

Dysfunction of the immune system can cause autoimmune diseases, inflammatory diseases and cancer. Immunodeficiency occurs when the immune system is less active than normal, resulting in recurring and life-threatening infections. In humans, immunodeficiency can be the result of a genetic disease such as severe combined immunodeficiency, acquired conditions such as HIV/AIDS, or the use of immunosuppressive medication. Autoimmunity results from a hyperactive immune system attacking normal tissues as if they were foreign organisms. Common autoimmune diseases include Hashimoto's thyroiditis, rheumatoid arthritis, diabetes mellitus type 1, and systemic lupus erythematosus. Immunology covers the study of all aspects of the immune system.

9.1.1 History of Immunology

Immunology is a science that examines the structure and function of the immune system. It originates from medicine and early studies on the causes of immunity to disease. The earliest known reference to immunity was during the plague of Athens in 430 BC. Thucydides noted that people who had recovered from a previous bout of the disease could nurse the sick without contracting the illness a second time. In the 18th century, Pierre-Louis Moreau de Maupertuis experimented with scorpion venom and observed that certain dogs and mice were immune to this venom. In the 10th century, Persian physician al-Razi (also known as Rhazes) wrote the first recorded theory of acquired immunity, noting that a smallpox bout protected its survivors from future infections. Although he explained the immunity in terms of "excess moisture" being expelled from the blood—therefore preventing a second occurrence of the disease—this theory explained many observations about smallpox known during this time.

These and other observations of acquired immunity were later exploited by Louis Pasteur in his development of vaccination and his proposed germ theory of disease. Pasteur's theory was in direct opposition to contemporary theories of disease, such as the miasma theory. It was not until Robert Koch's 1891 proofs, for which he was awarded a Nobel Prize in 1905, that microorganisms were confirmed as the cause of infectious disease. Viruses were confirmed as human pathogens in 1901, with the discovery of the yellow fever virus by Walter Reed.

Immunology made a great advance towards the end of the 19th century, through rapid developments in the study of humoral immunity and cellular immunity. Particularly important was the work of Paul Ehrlich, who proposed the side-chain theory to explain the specificity of the antigen-antibody reaction; his contributions to the understanding of humoral immunity were recognized by the award of a joint Nobel Prize in 1908, along with the founder of cellular immunology, Elie Metchnikoff. In 1974, Niels Kaj Jerne developed the immune network theory; he shared a Nobel Prize in 1984 with Georges J. F. Köhler and César Milstein for theories related to the immune system.

9.1.2 Layered Defense

The immune system protects its host from infection with layered defenses of increasing specificity. Physical barriers prevent pathogens such as bacteria and viruses from entering the organism. If a pathogen breaches these barriers, the innate immune system provides an immediate, but non-specific response. Innate immune systems are found in all animals. If pathogens successfully evade the innate response, vertebrates possess a second layer of protection, the adaptive immune system, which is activated by the innate response. Here, the immune system adapts its response during an infection to improve its recognition of the pathogen. This improved response is then retained after the pathogen has been eliminated, in the form of an immunological memory, and allows the adaptive immune system to mount faster and stronger attacks each time this pathogen is encountered.

Table 1: Components of the Immune System

Innate immune system	Adaptive immune system
Response is non-specific	Pathogen and antigen specific response
Exposure leads to immediate maximal response	Lag time between exposure and maximal response
Cell-mediated and humoral components	Cell-mediated and humoral components
No immunological memory	Exposure leads to immunological memory
Found in nearly all forms of life	Found only in jawed vertebrates

Both innate and adaptive immunity depend on the ability of the immune system to distinguish between self and non-self molecules. In immunology, *self* molecules are components of an organism's body that can be distinguished from foreign substances by the immune system. Conversely, *non-self* molecules are those recognized as foreign molecules. One class of non-self molecules are called antigens (originally named for being *anti*body *gen*erators) and are defined as substances that bind to specific immune receptors and elicit an immune response.

9.1.3 Surface Barriers

Several barriers protect organisms from infection, including mechanical, chemical, and biological barriers. The waxy cuticle of most leaves, the exoskeleton of insects, the shells and membranes of externally deposited eggs, and skin are examples of mechanical barriers that are the first line of defense against infection. Organisms cannot be completely sealed from their environments, so systems act to protect body openings such as the lungs, intestines, and the genitourinary tract. In the lungs, coughing and sneezing mechanically eject pathogens and other irritants from the respiratory tract. The flushing action of tears and urine also mechanically expels pathogens, while mucus secreted by the respiratory and gastrointestinal tract serves to trap and entangle microorganisms.

Chemical barriers also protect against infection. The skin and respiratory tract secrete antimicrobial peptides such as the β-defensins. Enzymes such as lysozyme and phospholipase A2 in saliva, tears, and breast milk are also antibacterials.

Vaginal secretions serve as a chemical barrier following menarche, when they become slightly acidic, while semen contains defensins and zinc to kill pathogens. In the stomach, gastric acid serves as a chemical defense against ingested pathogens.

Within the genitourinary and gastrointestinal tracts, commensal flora serve as biological barriers by competing with pathogenic bacteria for food and space and, in some cases, changing the conditions in their environment, such as pH or available iron. As a result, the probability that pathogens will reach sufficient numbers to cause illness is reduced.

9.1.4 Innate Immune System

Microorganisms or toxins that successfully enter an organism encounter the cells and mechanisms of the innate immune system. The innate response is usually triggered when microbes are identified by pattern recognition receptors, which recognize components that are conserved among broad groups of microorganisms, or when damaged, injured or stressed cells send out alarm signals, many of which are recognized by the same receptors as those that recognize pathogens. Innate immune defenses are non-specific, meaning these systems respond to pathogens in a generic way. This system does not confer long-lasting immunity against a pathogen. The innate immune system is the dominant system of host defense in most organisms, and the only one in plants.

Immune Sensing

Cells in the innate immune system use pattern recognition receptors to recognize molecular structures that are produced by pathogens. They are proteins expressed, mainly, by cells of the innate immune system, such as dendritic cells, macrophages, monocytes, neutrophils, and epithelial cells, to identify two classes of molecules: pathogen-associated molecular patterns (PAMPs), which are associated with microbial pathogens, and damage-associated molecular patterns (DAMPs), which are associated with components of host's cells that are released during cell damage or cell death.

Recognition of extracellular or endosomal PAMPs is mediated by transmembrane proteins known as toll-like receptors (TLRs). TLRs share a typical structural motif, the leucine rich repeats (LRRs), which give them a curved shape. Toll-like receptors were first discovered in *Drosophila* and trigger the synthesis and secretion of cytokines and activation of other host defense programs that are necessary for both innate or adaptive immune responses. Ten toll-like receptors have been described in humans.

Cells in the innate immune system have pattern recognition receptors, which detect infection or cell damage, inside. Three major classes of these "cytosolic" receptors are NOD–like receptors, RIG (retinoic acid-inducible gene)-like receptors, and cytosolic DNA sensors.

Innate Immune Cells

Some leukocytes (white blood cells) act like independent, single-celled organisms and are the second arm of the innate immune system. The innate leukocytes include the "professional" phagocytes (macrophages, neutrophils, and dendritic cells). These cells identify and eliminate pathogens, either by attacking larger pathogens through contact or by engulfing and then killing microorganisms. The other cells involved in the innate response include innate lymphoid cells, mast cells, eosinophils, basophils, and natural killer cells.

Phagocytosis is an important feature of cellular innate immunity performed by cells called phagocytes that engulf pathogens or particles. Phagocytes generally patrol the body searching for pathogens, but can be called to specific locations by cytokines. Once a pathogen has been engulfed by a phagocyte, it becomes trapped in an intracellular vesicle called a phagosome, which subsequently fuses with another vesicle called a lysosome to form a phagolysosome. The pathogen is killed by the activity of digestive enzymes or following a respiratory burst that releases free radicals into the phagolysosome. Phagocytosis evolved as a means of acquiring nutrients, but this role was extended in phagocytes to include engulfment of pathogens as a defense mechanism. Phagocytosis probably represents the oldest form of host defense, as phagocytes have been identified in both vertebrate and invertebrate animals.

Neutrophils and macrophages are phagocytes that travel throughout the body in pursuit of invading pathogens. Neutrophils are normally found in the bloodstream and are the most abundant type of phagocyte, representing 50% to 60% of total circulating leukocytes. During the acute phase of inflammation, neutrophils migrate toward the site of inflammation in a process called chemotaxis and are usually the first cells to arrive at the scene of infection. Macrophages are versatile cells that reside within tissues and produce an array of chemicals including enzymes, complement proteins, and cytokines. They can also act as scavengers that rid the body of worn-out cells and other debris and as antigen-presenting cells (APCs) that activate the adaptive immune system.

Dendritic cells are phagocytes in tissues that are in contact with the external environment; therefore, they are located mainly in the skin, nose, lungs, stomach, and intestines. They are named for their resemblance to neuronal dendrites, as both have many spine-like projections. Dendritic cells serve as a link between the bodily tissues and the innate and adaptive immune systems, as they present antigens to T cells, one of the key cell types of the adaptive immune system.

Granulocytes are leukocytes that have granules in their cytoplasm. In this category are neutrophils, mast cells, basophils, and eosinophils. Mast cells reside in connective tissues and mucous membranes and regulate the inflammatory response. They are most often associated with allergy and anaphylaxis. Basophils and eosinophils are related to neutrophils. They secrete chemical mediators that are involved in defending against parasites and play a role in allergic reactions, such as asthma. Innate lymphoid cells (ILCs) are a group of innate immune cells that are derived from common lymphoid progenitor and belong to the lymphoid lineage. These cells are defined by the absence of antigen-specific B- or T-cell receptor (TCR) because of the lack of recombination activating gene. ILCs do not express myeloid or dendritic cell markers. Natural killer cells (NK cells) are lymphocytes and a component of the innate immune system that does not directly attack invading microbes. Rather, NK cells destroy compromised host cells, such as tumor cells or virus-infected cells, recognizing such cells by a condition known as "missing self". This term describes cells with low levels of a cell-surface marker called MHC I (major histocompatibility complex)—a situation that can arise in viral infections of host cells. Normal body cells are not recognized and attacked by NK cells because they express intact self MHC antigens. Those MHC antigens are recognized by killer cell immunoglobulin receptors, which essentially put the brakes on NK cells.

Inflammation

Inflammation is one of the first responses of the immune system to infection. The symptoms of inflammation are redness, swelling, heat, and pain, which are caused by increased blood flow into tissue. Inflammation is produced by eicosanoids and cytokines, which are released by injured or infected cells. Eicosanoids include prostaglandins that produce fever and the dilation of blood vessels associated with inflammation and leukotrienes that attract certain white blood cells (leukocytes). Common cytokines include interleukins that are responsible for communication between white blood cells; chemokines that promote chemotaxis; and interferons that have antiviral effects, such as shutting down protein synthesis in the host cell. Growth factors and cytotoxic factors may also be released. These cytokines and other chemicals recruit immune cells to the site of infection and promote the healing of any damaged tissue following the removal of pathogens. The pattern-recognition receptors called inflammasomes are multiprotein complexes (consisting of an NLR, the adaptor protein ASC, and the effector molecule pro-caspase-1) that form in response to cytosolic PAMPs and DAMPs, whose function is to generate active forms of the inflammatory cytokines IL-1β and IL-18.

Humoral Defenses

The complement system is a biochemical cascade that attacks the surfaces of foreign cells. It contains over 20 different proteins and is named for its ability to "complement" the killing of pathogens by antibodies. Complement is the major humoral component of the innate immune response. Many species have complement systems, including non-mammals like plants, fish, and some invertebrates. In humans, this response is activated by complement binding to antibodies that have attached to these microbes or the binding of complement proteins to carbohydrates on the surfaces of microbes. This recognition signal triggers a rapid killing response.

The speed of the response is a result of signal amplification that occurs after sequential proteolytic activation of complement molecules, which are also proteases. After complement proteins initially bind to the microbe, they activate their protease activity, which in turn activates other complement proteases, and so on. This produces a catalytic cascade that amplifies the initial signal by controlled positive feedback. The cascade results in the production of peptides that attract immune cells, increase vascular permeability, and opsonize (coat) the surface of a pathogen, marking it for destruction. This deposition of complement can also kill cells directly by disrupting their plasma membrane via the formation of a membrane attack complex.

9.1.5 Adaptive Immune System

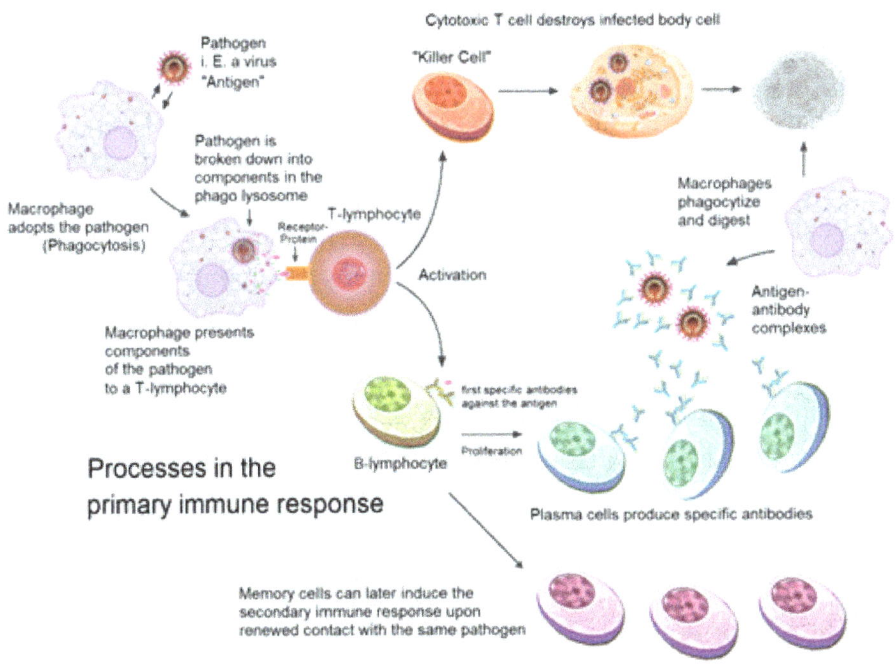

Fig 1: Overview of the Processes Involved in the Primary Immune Response

The adaptive immune system evolved in early vertebrates and allows for a stronger immune response as well as immunological memory, where each pathogen is "remembered" by a signature antigen. The adaptive immune response is antigen-specific and requires the recognition of specific "non-self" antigens during a process called antigen presentation. Antigen specificity allows for the generation of responses that are tailored to specific pathogens or pathogen-infected cells. The ability to mount these tailored responses is maintained in the body by "memory cells". Should a pathogen infect the body more than once, these specific memory cells are used to quickly eliminate it.

Recognition of Antigen

The cells of the adaptive immune system are special types of leukocytes, called lymphocytes. B cells and T cells are the major types of lymphocytes and are derived from hematopoietic stem cells in the bone marrow. B cells are involved in the humoral immune response, whereas T cells are involved in cell-mediated immune response. Killer T cells only recognize antigens coupled to Class I MHC molecules, while helper T cells and regulatory T cells only recognize antigens coupled to Class II MHC molecules. These two mechanisms of antigen presentation reflect the different roles of the two types of T cell. A third, minor subtype are the $\gamma\delta$ T cells that recognize intact antigens that are not bound to MHC receptors. The double-positive T cells are exposed to a wide variety of self-antigens in the thymus, in which iodine is necessary for its thymus development and activity. In contrast, the B cell antigen-specific receptor is an antibody molecule on the B cell surface and recognizes native (unprocessed) antigen without any need for antigen processing. Such antigens may be large molecules found on the surfaces of pathogens, but can also be small haptens (such as penicillin) attached to carrier molecule. Each lineage of B cell expresses a different antibody, so the complete set of B cell antigen receptors represent all the antibodies that the body can manufacture. When B or T cells encounter their related antigens they multiply and many "clones" of the cells are produced that target the same antigen. This is called clonal selection.

Antigen Presentation to T Lymphocytes

Both B cells and T cells carry receptor molecules that recognize specific targets. T cells recognize a "non-self" target, such as a pathogen, only after antigens (small fragments of the pathogen) have been processed and presented in combination with a "self" receptor called a major histocompatibility complex (MHC) molecule.

Cell Mediated Immunity

There are two major subtypes of T cells: the killer T cell and the helper T cell. In addition there are regulatory T cells which have a role in modulating immune response.

Killer T Cells

Killer T cells are a sub-group of T cells that kill cells that are infected with viruses (and other pathogens), or are otherwise damaged or dysfunctional. As with B cells, each type of T cell recognizes a different antigen. Killer T cells are activated when their T-cell receptor binds to this specific antigen in a complex with the MHC Class I receptor of another cell. Recognition of this MHC:antigen complex is aided by a co-receptor on the T cell, called CD8. The T cell then travels throughout the body in search of cells where the MHC I receptors bear this antigen. When an activated T cell contacts such cells, it releases cytotoxins, such as perforin, which form pores in the target cell's plasma membrane, allowing ions, water and toxins to enter. The entry of another toxin called granulysin (a protease) induces the target cell to undergo apoptosis. T cell killing of host cells is particularly important in preventing the replication of viruses. T cell activation is tightly controlled and generally requires a very strong MHC/antigen activation signal, or additional activation signals provided by "helper" T cells.

Helper T Cells

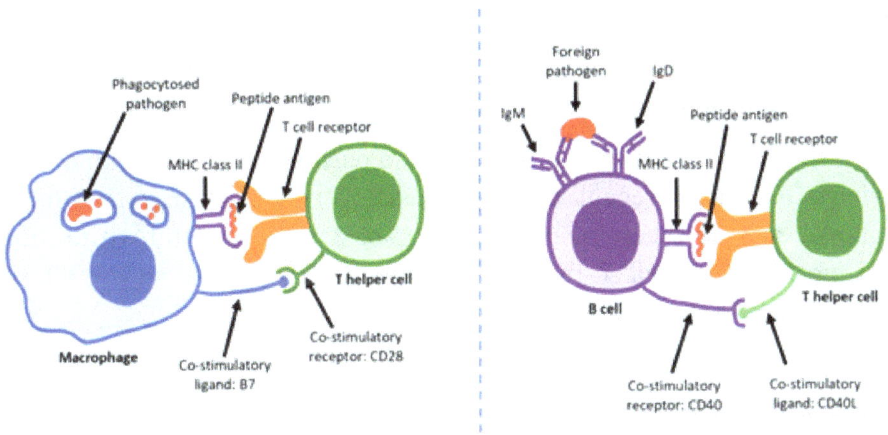

Fig 2: Activation of Macrophage or B Cell by T Helper Cell

Helper T cells regulate both the innate and adaptive immune responses and help determine which immune responses the body makes to a particular pathogen. These cells have no cytotoxic activity and do not kill infected cells or clear pathogens directly. They instead control the immune response by directing other cells to perform these tasks.

Helper T cells express T cell receptors that recognize antigen bound to Class II MHC molecules. The MHC:antigen complex is also recognized by the helper cell's CD4 co-receptor, which recruits molecules inside the T cell (such as Lck) that are responsible for the T cell's activation. Helper T cells have a weaker association with the MHC:antigen complex than observed for killer T cells, meaning many receptors (around 200–300) on the helper T cell must be bound by an MHC:antigen to activate the helper cell, while killer T cells can be activated by engagement of a single MHC:antigen molecule. Helper T cell activation also requires longer duration of engagement with an antigen-presenting cell.

The activation of a resting helper T cell causes it to release cytokines that influence the activity of many cell types. Cytokine signals produced by helper T cells enhance the microbicidal function of macrophages and the activity of killer T cells. In addition, helper T cell activation causes an upregulation of molecules expressed on the T cell's surface, such as CD40 ligand (also called CD154), which provide extra stimulatory signals typically required to activate antibody-producing B cells.

Gamma Delta T Cells

Gamma delta T cells (γ δ T cells) possess an alternative T-cell receptor (TCR) as opposed to CD4+ and CD8+ (α β) T cells and share the characteristics of helper T cells, cytotoxic T cells and NK cells. The conditions that produce responses from γ δ T cells are not fully understood. Like other 'unconventional' T cell subsets bearing invariant TCRs, such as CD1d-restricted natural killer T cells, γ δ T cells straddle the border between innate and adaptive immunity.

On one hand, γ δ T cells are a component of adaptive immunity as they rearrange TCR genes to produce receptor diversity and can also develop a memory phenotype.

On the other hand, the various subsets are also part of the innate immune system, as restricted TCR or NK receptors may be used as pattern recognition receptors. For example, large numbers of human Vγ9/Vδ2 T cells respond within hours to common molecules produced by microbes, and highly restricted Vδ1+ T cells in epithelia respond to stressed epithelial cells.

Humoral Immune Response

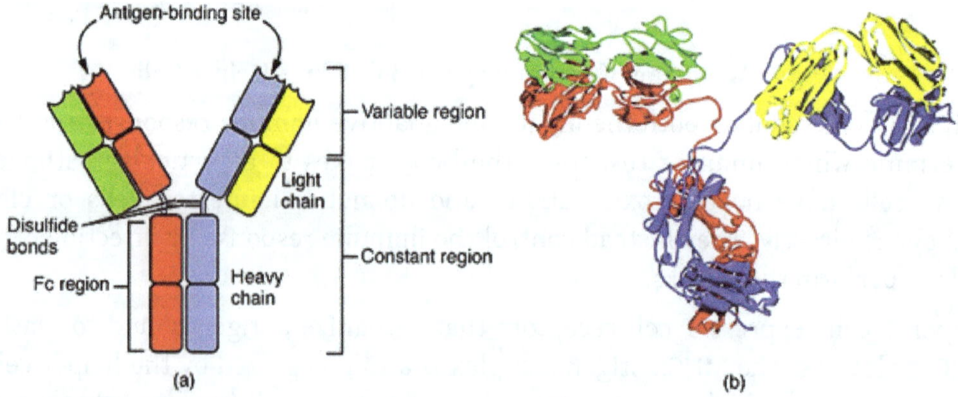

Fig 3: An Antibody is Made Up of Two Heavy Chains and Two Light Chains. The Unique Variable Region Allows an Antibody to Recognize its Matching Antigen

A B cell identifies pathogens when antibodies on its surface bind to a specific foreign antigen. This antigen/antibody complex is taken up by the B cell and processed by proteolysis into peptides. The B cell then displays these antigenic peptides on its surface MHC class II molecules. This combination of MHC and antigen attracts a matching helper T cell, which releases lymphokines and activates the B cell. As the activated B cell then begins to divide, its offspring (plasma cells) secrete millions of copies of the antibody that recognizes this antigen. These antibodies circulate in blood plasma and lymph, bind to pathogens expressing the antigen and mark them for destruction by complement activation or for uptake and destruction by phagocytes. Antibodies can also neutralize challenges directly, by binding to bacterial toxins or by interfering with the receptors that viruses and bacteria use to infect cells.

Newborn infants have no prior exposure to microbes and are particularly vulnerable to infection. Several layers of passive protection are provided by the mother. During pregnancy, a particular type of antibody, called IgG, is transported from mother to baby directly through the placenta, so human babies have high levels of antibodies even at birth, with the same range of antigen specificities as their mother. Breast milk or colostrum also contains antibodies that are transferred to the gut of the infant and protect against bacterial infections until the newborn can synthesize its own antibodies. This is passive immunity because the fetus does not actually make any memory cells or antibodies—it only borrows them. This passive immunity is usually short-term, lasting from a few days up to several months. In medicine, protective passive immunity can also be transferred artificially from one individual to another.

Immunological Memory

When B cells and T cells are activated and begin to replicate, some of their offspring become long-lived memory cells. Throughout the lifetime of an animal, these memory cells remember each specific pathogen encountered and can mount a strong response if the pathogen is detected again. T-cells recognize pathogens by small protein-based infection signals, called antigens, that bind to directly to T-cell surface receptors. B-cells use the protein, immunoglobulin, to recognize pathogens by their antigens. This is "adaptive" because it occurs during the lifetime of an individual as an adaptation to infection with that pathogen and prepares the immune system for future challenges. Immunological memory can be in the form of either passive short-term memory or active long-term memory.

9.1.6 Physiological Regulation

The immune system is involved in many aspects of physiological regulation in the body. The immune system interacts intimately with other systems, such as the endocrine and the nervous systems. The immune system also plays a crucial role in embryogenesis (development of the embryo), as well as in tissue repair and regeneration.

Hormones

Hormones can act as immunomodulators, altering the sensitivity of the immune system. For example, female sex hormones are known immunostimulators of both adaptive and innate immune responses. Some autoimmune diseases such as lupus erythematosus strike women preferentially, and their onset often coincides with puberty. By contrast, male sex hormones such as testosterone seem to be immunosuppressive. Other hormones appear to regulate the immune system as well, most notably prolactin, growth hormone and vitamin D.

Vitamin D

Although cellular studies indicate that vitamin D has receptors and probable functions in the immune system, there is no clinical evidence to prove that vitamin D deficiency increases the risk for immune diseases or vitamin D supplementation lowers immune disease risk. A 2011 United States Institute of Medicine report stated that "outcomes related to ... immune functioning and autoimmune disorders, and infections ... could not be linked reliably with calcium or vitamin D intake and were often conflicting."

Sleep and Rest

The immune system is affected by sleep and rest, and sleep deprivation is detrimental to immune function. Complex feedback loops involving cytokines, such as interleukin-1 and tumor necrosis factor-α produced in response to infection, appear to also play a role in the regulation of non-rapid eye movement (REM) sleep. Thus the immune response to infection may result in changes to the sleep cycle, including an increase in slow-wave sleep relative to REM sleep.

In people with sleep deprivation, active immunizations may have a diminished effect and may result in lower antibody production, and a lower immune response, than would be noted in a well-rested individual. Additionally, proteins such as NFIL3, which have been shown to be closely intertwined with both T-cell differentiation and circadian rhythms, can be affected through the disturbance of natural light and dark cycles through instances of sleep deprivation. These disruptions can lead to an increase in chronic conditions such as heart disease, chronic pain, and asthma.

In addition to the negative consequences of sleep deprivation, sleep and the intertwined circadian system have been shown to have strong regulatory effects on immunological functions affecting both innate and adaptive immunity. First, during the early slow-wave-sleep stage, a sudden drop in blood levels of cortisol, epinephrine, and norepinephrine causes increased blood levels of the hormones leptin, pituitary growth hormone, and prolactin. These signals induce a pro-inflammatory state through the production of the pro-inflammatory cytokines interleukin-1, interleukin-12, TNF-alpha and IFN-gamma. These cytokines then stimulate immune functions such as immune cell activation, proliferation, and differentiation. During this time of a slowly evolving adaptive immune response, there is a peak in undifferentiated or less differentiated cells, like naïve and central memory T cells. In addition to these effects, the milieu of hormones produced at this time (leptin, pituitary growth hormone, and prolactin) supports the interactions between APCs and T-cells, a shift of the T_h1/T_h2 cytokine balance towards one that supports T_h1, an increase in overall T_h cell proliferation, and naïve T cell migration to lymph nodes. This is also thought to support the formation of long-lasting immune memory through the initiation of Th1 immune responses.

During wake periods, differentiated effector cells, such as cytotoxic natural killer cells and cytotoxic T lymphocytes, peak to elicit an effective response against any intruding pathogens. Anti-inflammatory molecules, such as cortisol and catecholamines, also peak during awake active times. Inflammation would cause serious cognitive and physical impairments if it were to occur during wake times, and inflammation may occur during sleep times due to the presence of melatonin. Inflammation causes a great deal of oxidative stress and the presence of melatonin during sleep times could actively counteract free radical production during this time.

Physical Exercise

Physical exercise has a positive effect on the immune system and depending on the frequency and intensity, the pathogenic effects of diseases caused by bacteria and viruses are moderated. Immediately after intense exercise there is a transient immunodepression, where the number of circulating lymphocytes decreases and antibody production declines. This may give rise to a window of opportunity for infection and reactivation of latent virus infections, but the evidence is inconclusive.

Changes at the Cellular Level

During exercise there is an increase in circulating white blood cells of all types. This is caused by the frictional force of blood flowing on the endothelial cell surface and catecholamines affecting β-adrenergic receptors (βARs). The number of neutrophils in the blood increases and remains raised for up to six hours and immature forms are present.

Although the increase in neutrophils ("neutrophilia") is similar to that seen during bacterial infections, after exercise the cell population returns to normal by around 24 hours.

The number of circulating lymphocytes (mainly natural killer cells) decreases during intense exercise but returns to normal after 4 to 6 hours. Although up to 2% of the cells die most migrate from the blood to the tissues, mainly the intestines and lungs, where pathogens are most likely to be encountered.

Some monocytes leave the blood circulation and migrate to the muscles where they differentiate and become macrophages. These cells differentiate into two types: proliferative macrophages, which are responsible for increasing the number of stem cells and restorative macrophages, which are involved their maturing to muscle cells.

Repair and Regeneration

The immune system, particularly the innate component, plays a decisive role in tissue repair after an insult. Key actors include macrophages and neutrophils, but other cellular actors, including γ δ T cells, innate lymphoid cells (ILCs), and regulatory T cells (Tregs), are also important. The plasticity of immune cells and the balance between pro-inflammatory and anti-inflammatory signals are crucial aspects of efficient tissue repair. Immune components and pathways are involved in regeneration as well, for example in amphibians such as in axolotl limb regeneration. According to one hypothesis, organisms that can regenerate (*e.g.*, axolotls) could be less immunocompetent than organisms that cannot regenerate.

9.1.7 Disorders of Human Immunity

Immunodeficiencies

Immunodeficiencies occur when one or more of the components of the immune system are inactive. The ability of the immune system to respond to pathogens is diminished in both the young and the elderly, with immune responses beginning to decline at around 50 years of age due to immunosenescence. In developed countries, obesity, alcoholism, and drug use are common causes of poor immune function, while malnutrition is the most common cause of immunodeficiency in developing countries. Diets lacking sufficient protein are associated with impaired cell-mediated immunity, complement activity, phagocyte function, IgA antibody concentrations, and cytokine production. Additionally, the loss of the thymus at an early age through genetic mutation or surgical removal results in severe immunodeficiency and a high susceptibility to infection. Immunodeficiencies can also be inherited or 'acquired'. Severe combined immunodeficiency is a rare genetic disorder characterized by the disturbed development of functional T cells and B cells caused by numerous genetic mutations. Chronic granulomatous disease, where phagocytes have a reduced ability to destroy pathogens, is an example of an inherited, or congenital, immunodeficiency. AIDS and some types of cancer cause acquired immunodeficiency.

Autoimmunity

Overactive immune responses form the other end of immune dysfunction, particularly the autoimmune diseases. Here, the immune system fails to properly distinguish between self and non-self, and attacks part of the body. Under normal circumstances, many T cells and antibodies react with "self" peptides. One of the functions of specialized cells (located in the thymus and bone marrow) is to present young lymphocytes with self antigens produced throughout the body and to eliminate those cells that recognize self-antigens, preventing autoimmunity. Common autoimmune diseases include Hashimoto's thyroiditis, rheumatoid arthritis, diabetes mellitus type 1, and systemic lupus erythematosus.

Hypersensitivity

Hypersensitivity is an immune response that damages the body's own tissues. It is divided into four classes (Type I – IV) based on the mechanisms involved and the time course of the hypersensitive reaction. Type I hypersensitivity is an immediate or anaphylactic reaction, often associated with allergy. Symptoms can range from mild discomfort to death. Type I hypersensitivity is mediated by IgE, which triggers degranulation of mast cells and basophils when cross-linked by antigen. Type II hypersensitivity occurs when antibodies bind to antigens on the individual's own cells, marking them for destruction. This is also called antibody-dependent (or cytotoxic) hypersensitivity, and is mediated by IgG and IgM antibodies. Immune complexes (aggregations of antigens, complement proteins, and IgG and IgM antibodies) deposited in various tissues trigger Type III hypersensitivity reactions. Type IV hypersensitivity (also known as cell-mediated or *delayed type hypersensitivity*) usually takes between two and three days to develop. Type IV reactions are involved in many autoimmune and infectious diseases, but may also involve contact dermatitis. These reactions are mediated by T cells, monocytes, and macrophages.

Idiopathic Inflammation

Inflammation is one of the first responses of the immune system to infection, but it can appear without known cause. Inflammation is produced by eicosanoids and cytokines, which are released by injured or infected cells. Eicosanoids include prostaglandins that produce fever and the dilation of blood vessels associated with inflammation, and leukotrienes that attract certain white blood cells (leukocytes). Common cytokines include interleukins that are responsible for communication between white blood cells; chemokines that promote chemotaxis; and interferons that have anti-viral effects, such as shutting down protein synthesis in the host cell. Growth factors and cytotoxic factors may also be released. These cytokines and other chemicals recruit immune cells to the site of infection and promote healing of any damaged tissue following the removal of pathogens.

9.1.8 Manipulation in Medicine

The immune response can be manipulated to suppress unwanted responses resulting from autoimmunity, allergy, and transplant rejection, and to stimulate protective responses against pathogens that largely elude the immune system or cancer.

Immunosuppression

Immunosuppressive drugs are used to control autoimmune disorders or inflammation when excessive tissue damage occurs, and to prevent rejection after an organ transplant.

Anti-inflammatory drugs are often used to control the effects of inflammation. Glucocorticoids are the most powerful of these drugs and can have many undesirable side effects, such as central obesity, hyperglycemia, and osteoporosis. Their use is tightly controlled. Lower doses of anti-inflammatory drugs are often used in conjunction with cytotoxic or immunosuppressive drugs such as methotrexate or azathioprine.

Cytotoxic drugs inhibit the immune response by killing dividing cells such as activated T cells. This killing is indiscriminate and other constantly dividing cells and their organs are affected, which causes toxic side effects. Immunosuppressive drugs such as cyclosporin prevent T cells from responding to signals correctly by inhibiting signal transduction pathways.

Immunostimulation

Claims made by marketers of various products and alternative health providers, such as chiropractors, homeopaths, and acupuncturists to be able to stimulate or "boost" the immune system generally lack meaningful explanation and evidence of effectiveness.

Vaccination

Long-term *active* memory is acquired following infection by activation of B and T cells. Active immunity can also be generated artificially, through vaccination. The principle behind vaccination (also called immunization) is to introduce an antigen from a pathogen to stimulate the immune system and develop specific immunity against that particular pathogen without causing disease associated with that organism.

This deliberate induction of an immune response is successful because it exploits the natural specificity of the immune system, as well as its inducibility. With infectious disease remaining one of the leading causes of death in the human population, vaccination represents the most effective manipulation of the immune system mankind has developed.

Many vaccines are based on acellular components of micro-organisms, including harmless toxin components. Since many antigens derived from acellular vaccines do not strongly induce the adaptive response, most bacterial vaccines are provided with additional adjuvants that activate the antigen-presenting cells of the innate immune system and maximize immunogenicity.

Tumor Immunology

Another important role of the immune system is to identify and eliminate tumors. This is called immune surveillance. The *transformed cells* of tumors express antigens that are not found on normal cells. To the immune system, these antigens appear foreign, and their presence causes immune cells to attack the transformed tumor cells. The antigens expressed by tumors have several sources; some are derived from oncogenic viruses like human papillomavirus, which causes cancer of the cervix, vulva, vagina, penis, anus, mouth, and throat, while others are the organism's own proteins that occur at low levels in normal cells but reach high levels in tumor cells. One example is an enzyme called tyrosinase that, when expressed at high levels, transforms certain skin cells (for example, melanocytes) into tumors called melanomas. A third possible source of tumor antigens are proteins normally important for regulating cell growth and survival, that commonly mutate into cancer inducing molecules called oncogenes.

The main response of the immune system to tumors is to destroy the abnormal cells using killer T cells, sometimes with the assistance of helper T cells. Tumor antigens are presented on MHC class I molecules in a similar way to viral antigens. This allows killer T cells to recognize the tumor cell as abnormal. NK cells also kill tumorous cells in a similar way, especially if the tumor cells have fewer MHC class I molecules on their surface than normal; this is a common phenomenon with tumors. Sometimes antibodies are generated against tumor cells allowing for their destruction by the complement system.

Some tumors evade the immune system and go on to become cancers. Tumor cells often have a reduced number of MHC class I molecules on their surface, thus avoiding detection by killer T cells. Some tumor cells also release products that inhibit the immune response; for example by secreting the cytokine TGF-β, which suppresses the activity of macrophages and lymphocytes. In addition, immunological tolerance may develop against tumor antigens, so the immune system no longer attacks the tumor cells.

Paradoxically, macrophages can promote tumor growth when tumor cells send out cytokines that attract macrophages, which then generate cytokines and growth factors such as tumor-necrosis factor alpha that nurture tumor development or promote stem-cell-like plasticity. In addition, a combination of hypoxia in the tumor and a cytokine produced by macrophages induces tumor cells to decrease production of a protein that blocks metastasis and thereby assists spread of cancer cells. Anti-tumor M1 macrophages are recruited in early phases to tumor development but are progressively differentiated to M2 with pro-tumor effect, an immunosuppressor switch. The hypoxia reduces the cytokine production for the anti-tumor response and progressively macrophages acquire pro-tumor M2 functions driven by the tumor microenvironment, including IL-4 and IL-10. Cancer immunotherapy covers the medical ways to stimulate the immune system to attack cancer tumors.

Predicting Immunogenicity

Some drugs can cause a neutralizing immune response, meaning that the immune system produces neutralizing antibodies that counteract the action of the drugs, particularly if the drugs are administered repeatedly, or in larger doses. This limits the effectiveness of drugs based on larger peptides and proteins (which are typically larger than 6000 Da). In some cases, the drug itself is not immunogenic, but may be co-administered with an immunogenic compound, as is sometimes the case for Taxol. Computational methods have been developed to predict the immunogenicity of peptides and proteins, which are particularly useful in designing therapeutic antibodies, assessing likely virulence of mutations in viral coat particles, and validation of proposed peptide-based drug treatments. Early techniques relied mainly on the observation that hydrophilic amino acids are overrepresented in epitope regions than hydrophobic amino acids; however, more recent developments rely on machine learning techniques using databases of existing known epitopes, usually on well-studied virus proteins, as a training set. A publicly accessible database has been established for the cataloguing of epitopes from pathogens known to be recognizable by B cells. The emerging field of bioinformatics-based studies of immunogenicity is referred to as *immunoinformatics*. Immunoproteomics is the study of large sets of proteins (proteomics) involved in the immune response.

9.1.9 Evolution and other Mechanisms

Evolution of the Immune System

It is likely that a multicomponent, adaptive immune system arose with the first vertebrates, as invertebrates do not generate lymphocytes or an antibody-based humoral response. Immune systems evolved in deuterostomes as shown in the cladogram.

Many species, however, use mechanisms that appear to be precursors of these aspects of vertebrate immunity. Immune systems appear even in the structurally simplest forms of life, with bacteria using a unique defense mechanism, called the restriction modification system to protect themselves from viral pathogens, called bacteriophages. Prokaryotes (bacteria and archea) also possess acquired immunity, through a system that uses CRISPR sequences to retain fragments of the genomes of phage that they have come into contact with in the past, which allows them to block virus replication through a form of RNA interference. Prokaryotes also possess other defense mechanisms. Offensive elements of the immune systems are also present in unicellular eukaryotes, but studies of their roles in defense are few.

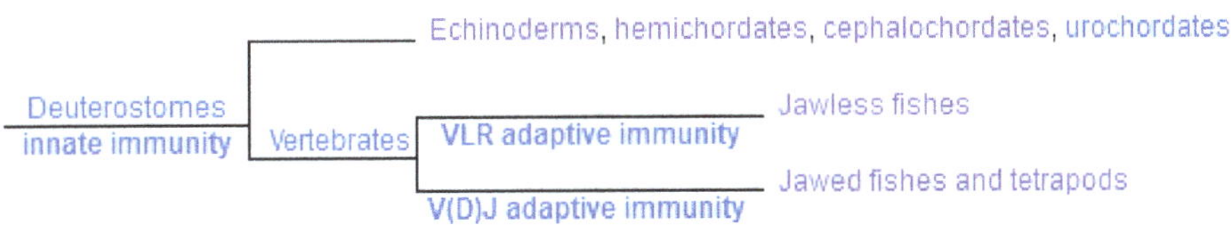

Pattern recognition receptors are proteins used by nearly all organisms to identify molecules associated with pathogens. Antimicrobial peptides called defensins are an evolutionarily conserved component of the innate immune response found in all animals and plants, and represent the main form of invertebrate systemic immunity. The complement system and phagocytic cells are also used by most forms of invertebrate life. Ribonucleases and the RNA interference pathway are conserved across all eukaryotes, and are thought to play a role in the immune response to viruses.

Unlike animals, plants lack phagocytic cells, but many plant immune responses involve systemic chemical signals that are sent through a plant. Individual plant cells respond to molecules associated with pathogens known as pathogen-associated molecular patterns or PAMPs.

When a part of a plant becomes infected, the plant produces a localized hypersensitive response, whereby cells at the site of infection undergo rapid apoptosis to prevent the spread of the disease to other parts of the plant. Systemic acquired resistance is a type of defensive response used by plants that renders the entire plant resistant to a particular infectious agent. RNA silencing mechanisms are particularly important in this systemic response as they can block virus replication.

Alternative Adaptive Immune System

Evolution of the adaptive immune system occurred in an ancestor of the jawed vertebrates. Many of the classical molecules of the adaptive immune system (for example, immunoglobulins and T-cell receptors) exist only in jawed vertebrates. A distinct lymphocyte-derived molecule has been discovered in primitive jawless vertebrates, such as the lamprey and hagfish. These animals possess a large array of molecules called Variable lymphocyte receptors (VLRs) that, like the antigen receptors of jawed vertebrates, are produced from only a small number (one or two) of genes. These molecules are believed to bind pathogenic antigens in a similar way to antibodies, and with the same degree of specificity.

Manipulation by Pathogens

The success of any pathogen depends on its ability to elude host immune responses. Therefore, pathogens evolved several methods that allow them to successfully infect a host, while evading detection or destruction by the immune system. Bacteria often overcome physical barriers by secreting enzymes that digest the barrier, for example, by using a type II secretion system. Alternatively, using a type III secretion system, they may insert a hollow tube into the host cell, providing a direct route for proteins to move from the pathogen to the host. These proteins are often used to shut down host defenses.

An evasion strategy used by several pathogens to avoid the innate immune system is to hide within the cells of their host (also called intracellular pathogenesis). Here, a pathogen spends most of its life-cycle inside host cells, where it is shielded from direct contact with immune cells, antibodies and complement. Some examples of intracellular pathogens include viruses, the food poisoning bacterium *Salmonella* and the eukaryotic parasites that cause malaria (*Plasmodium spp.*) and leishmaniasis (*Leishmania spp.*). Other bacteria, such as *Mycobacterium tuberculosis*, live inside a protective capsule that prevents lysis by complement. Many pathogens secrete compounds that diminish or misdirect the host's immune response. Some bacteria form biofilms to protect themselves from the cells and proteins of the immune system. Such biofilms are present in many successful infections, such as the chronic *Pseudomonas aeruginosa* and *Burkholderia cenocepacia* infections characteristic of cystic fibrosis. Other bacteria generate surface proteins that bind to antibodies, rendering them ineffective; examples include *Streptococcus* (protein G), *Staphylococcus aureus* (protein A), and *Peptostreptococcus magnus* (protein L).

The mechanisms used to evade the adaptive immune system are more complicated. The simplest approach is to rapidly change non-essential epitopes (amino acids and/or sugars) on the surface of the pathogen, while keeping essential epitopes concealed. This is called antigenic variation. An example is HIV, which mutates rapidly, so the proteins on its viral envelope that are essential for entry into its host target cell are constantly changing. These frequent changes in antigens may explain the failures of vaccines directed at this virus. The parasite *Trypanosoma brucei* uses a similar strategy, constantly switching one type of surface protein for another, allowing it to stay one step ahead of the antibody response. Masking antigens with host molecules is another common strategy for avoiding detection by the immune system. In HIV, the envelope that covers the virion is formed from the outermost membrane of the host cell; such "self-cloaked" viruses make it difficult for the immune system to identify them as "non-self" structures.

9.2 KEY COMPONENTS OF THE IMMUNE SYSTEM

The immune system is composed of a wide variety of cells, tissues, and organs, each playing a vital role in the body's defense mechanisms.

- **White Blood Cells (Leukocytes):** These cells are the cornerstone of the immune system. There are different types of white blood cells, including neutrophils, monocytes, lymphocytes (B cells, T cells, and natural killer cells), and eosinophils. Each of these cells has a specific function in detecting, targeting, and eliminating pathogens or abnormal cells.
  - *T Cells:* T cells are a type of lymphocyte that play a key role in cell-mediated immunity. They help to recognize and kill infected or cancerous cells. There are several subtypes of T cells, including cytotoxic T cells, which directly destroy infected cells, and helper T cells, which coordinate the immune response.
  - *B Cells:* B cells are responsible for producing antibodies. These antibodies can neutralize pathogens by binding to them, marking them for destruction by other immune cells. B cells can differentiate into plasma cells that secrete large amounts of antibodies during an immune response.
  - *Natural Killer (NK) Cells:* These cells are part of the innate immune system and are important in identifying and eliminating cells infected with viruses or transformed into cancer cells.
- **Lymphoid Organs:** The immune system also relies on primary and secondary lymphoid organs. Primary lymphoid organs include the bone marrow and thymus, where immune cells are produced and mature. Secondary lymphoid organs include the lymph nodes, spleen, and mucosal-associated lymphoid tissues (MALT), where immune cells encounter pathogens and initiate immune responses.
- **Antibodies:** Antibodies (or immunoglobulins) are proteins produced by B cells in response to specific pathogens. These proteins recognize and bind to antigens (foreign molecules) on the surface of pathogens, marking them for destruction by other immune cells.

9.3 IMMUNE RESPONSE AND MECHANISMS

The immune system uses a variety of mechanisms to detect and eliminate pathogens. These mechanisms involve both innate and adaptive components.

- **Phagocytosis:** One of the primary responses of the innate immune system is the process of phagocytosis, where immune cells such as neutrophils and macrophages engulf and digest pathogens. This is the body's way of removing pathogens directly.
- **Inflammation:** Inflammation is an essential part of the immune

response. It occurs when immune cells release signaling molecules like cytokines, which promote blood flow to the affected area, attract more immune cells, and increase the permeability of blood vessels to allow immune cells to enter tissues. Inflammation can be protective, but when excessive or chronic, it can contribute to disease.

- **Antigen Presentation**: In adaptive immunity, antigen-presenting cells (APCs), such as dendritic cells and macrophages, process pathogens and present fragments of these pathogens (antigens) on their surface. These antigens are recognized by T cells, which are activated to initiate a targeted immune response.
- **Memory Response**: Once the immune system has encountered a pathogen, it develops a memory of the pathogen's antigens. Memory B cells and T cells are generated during an infection, which allows the immune system to mount a faster and more effective response if the pathogen is encountered again.

9.4 TYPES OF IMMUNITY

Immunity can be divided into two major categories: **innate immunity** and **adaptive immunity**. Each plays a crucial role in the body's defense against infection.

- **Active Immunity**: Active immunity occurs when the body produces its own immune response against a pathogen, often through exposure to the pathogen or through vaccination. This can involve the production of antibodies and the activation of T cells. Active immunity is long-lasting and provides the body with immunological memory, which protects against future infections.
- **Passive Immunity**: Passive immunity is conferred when antibodies are transferred from one individual to another, either naturally (such as through the placenta from mother to child) or artificially (such as through antibody-based treatments or immune globulin therapy). Passive immunity provides immediate protection, but it is short-lived because the recipient's immune system does not produce its own antibodies.

9.5 IMMUNE SYSTEM DISORDERS

While the immune system is designed to protect the body, it can sometimes malfunction, leading to various disorders.

- **Autoimmune Diseases:** Autoimmune diseases occur when the immune system mistakenly attacks the body's own cells and tissues. Examples include rheumatoid arthritis, lupus, and multiple sclerosis. In these conditions, the immune system's inability to distinguish self from non-self results in chronic inflammation and tissue damage.

- **Allergies**: Allergies occur when the immune system overreacts to harmless substances, such as pollen, dust mites, or certain foods. This exaggerated immune response leads to inflammation and tissue damage. Common allergic conditions include asthma, hay fever, and food allergies.
- **Immunodeficiencies:** Immunodeficiency disorders occur when the immune system is weakened or absent, making individuals more susceptible to infections. Primary immunodeficiencies are typically genetic, while secondary immunodeficiencies can result from infections like HIV/AIDS, which specifically targets immune cells.

9.6 IMMUNOLOGY RESEARCH AND FUTURE DIRECTIONS

Immunology research continues to evolve, with significant advances in understanding the immune system and its role in disease. Current areas of focus in immunology include:

- **Cancer Immunotherapy**: Immunotherapies that harness the body's immune system to fight cancer, such as checkpoint inhibitors and CAR-T cell therapies, have shown promising results in treating various cancers.
- **Personalized Medicine**: Advances in genomics and immunology have paved the way for personalized medicine, where treatments are tailored to an individual's unique immune system and genetic makeup.
- **Immunological Memory**: Researchers are exploring ways to enhance immunological memory, not only for infectious diseases but also for cancer and chronic conditions.

9.7 CONCLUSION

Immunology is a rapidly evolving field that holds the key to understanding the body's defense mechanisms and developing new treatments for infectious diseases, autoimmune disorders, cancer, and allergies. By continuing to explore the intricacies of the immune system and harnessing its power, scientists and clinicians can develop more effective therapies, vaccines, and strategies to promote health and treat disease. With ongoing advancements in immunology, the future holds great promise for improving human health and combating the world's most challenging diseases.

CHAPTER 10
VACCINATION AND IMMUNIZATION

Vaccination and immunization have long been recognized as some of the most effective public health tools in preventing infectious diseases. They have not only saved millions of lives but have also significantly reduced the prevalence of several deadly diseases worldwide. The concepts of vaccination and immunization are fundamental in modern medicine, as they harness the body's natural defense mechanisms to prevent diseases before they occur.

10.1 UNDERSTANDING VACCINATION AND IMMUNIZATION

10.1.1 Vaccination

Vaccination is the administration of a vaccine to help the immune system develop immunity from a disease. Vaccines contain a microorganism or virus in a weakened, live or killed state, or proteins or toxins from the organism. In stimulating the body's adaptive immunity, they help prevent sickness from an infectious disease. When a sufficiently large percentage of a population has been vaccinated, herd immunity results. Herd immunity protects those who may be immunocompromised and cannot get a vaccine because even a weakened version would harm them. The effectiveness of vaccination has been widely studied and verified. Vaccination is the most effective method of preventing infectious diseases; widespread immunity due to vaccination is largely responsible for the worldwide eradication of smallpox and the elimination of diseases such as polio and tetanus from much of the world. However, some diseases, such as measles outbreaks in America, have seen rising cases due to relatively low vaccination rates in the 2010s – attributed, in part, to vaccine hesitancy. According to the World Health Organization, vaccination prevents 3.5–5 million deaths per year.

The first disease people tried to prevent by inoculation was most likely smallpox, with the first recorded use of variolation occurring in the 16th century in China. It was also the first disease for which a vaccine was produced. Although at least six people had used the same principles years earlier, the smallpox vaccine was invented in 1796 by English physician Edward Jenner. He was the first to publish evidence that it was effective and to provide advice on its production. Louis Pasteur furthered the concept through his work in microbiology. The immunization was called *vaccination* because it was derived from a virus affecting cows (Latin: *vacca* 'cow'). Smallpox was a contagious and deadly disease, causing the deaths of 20–60% of infected adults and over 80% of infected children. When smallpox was finally eradicated in 1979, it had already killed an estimated 300–500 million people in the 20th century.

Vaccination and *immunization* have a similar meaning in everyday language. This is distinct from inoculation, which uses unweakened live pathogens. Vaccination efforts have been met with some reluctance on scientific, ethical, political, medical safety, and religious grounds, although no major religions oppose vaccination, and some consider it an obligation due to the potential to save lives. In the United States, people may receive compensation for alleged injuries under the National Vaccine Injury Compensation Program. Early success brought widespread acceptance, and mass vaccination campaigns have greatly reduced the incidence of many diseases in numerous geographic regions. The Centers for Disease Control and Prevention lists vaccination as one of the ten great public health achievements of the 20th century in the U.S.

History

Before the first vaccinations, in the sense of using cowpox to inoculate people against smallpox, people have been inoculated in China and elsewhere, before being copied in the west, by using smallpox, called variolation. The earliest hints of the practice of variolation for smallpox in China come during the 10th century. The Chinese also practiced the oldest documented use of variolation, which comes from Wan Quan's (1499–1582) Douzhen Xinfa of 1549. They implemented a method of "nasal insufflation" administered by blowing powdered smallpox material, usually scabs, up the nostrils. Various insufflation techniques have been recorded throughout the sixteenth and seventeenth centuries within China. Two reports on the Chinese practice of inoculation were received by the Royal Society in London in 1700; one by Martin Lister who received a report by an employee of the East India Company stationed in China and another by Clopton Havers. In France, Voltaire reports that the Chinese have practiced variolation "these hundred years".

In 1796, Edward Jenner, a doctor in Berkeley in Gloucestershire, England, tested a common theory that a person who had contracted cowpox would be immune from smallpox. To test the theory, he took cowpox vesicles from a milkmaid named Sarah Nelmes with which he infected an eight-year-old boy named James Phipps, and two months later he inoculated the boy with smallpox, and smallpox did not develop. In 1798, Jenner published *An Inquiry Into the Causes and Effects of the Variolæ Vaccinæ* which created widespread interest. He distinguished 'true' and 'spurious' cowpox (which did not give the desired effect) and developed an "arm-to-arm" method of propagating the vaccine from the vaccinated individual's pustule. Early attempts at confirmation were confounded by contamination with smallpox, but despite controversy within the medical profession and religious opposition to the use of animal material, by 1801 his report was translated into six languages and over 100,000 people were vaccinated. The term *vaccination* was coined in 1800 by the surgeon Richard Dunning in his text *Some observations on vaccination.*

In 1802, the Scottish physician Helenus Scott vaccinated dozens of children in Bombay against smallpox using Jenner's cowpox vaccine. In the same year Scott penned a letter to the editor in the *Bombay Courier*, declaring that "We have it now in our power to communicate the benefits of this important discovery to every part of India, perhaps to China and the whole eastern world". Subsequently, vaccination became firmly established in British India. A vaccination campaign was started in the new British colony of Ceylon in 1803. By 1807 the British had vaccinated more than a million Indians and Sri Lankans against smallpox. Also in 1803 the Spanish Balmis Expedition launched the first transcontinental effort to vaccinate people against smallpox. Following a smallpox epidemic in 1816 the Kingdom of Nepal ordered smallpox vaccine and requested the English veterinarian William Moorcroft to help in launching a vaccination campaign. In the same year a law was passed in Sweden to require the vaccination of children against smallpox by the age of two. Prussia briefly introduced compulsory vaccination in 1810 and again in the 1920s, but decided against a compulsory vaccination law in 1829. A law on compulsory smallpox vaccination was introduced in the Province of Hanover in the 1820s. In 1826, in Kragujevac, future prince Mihailo of Serbia was the first person to be vaccinated against smallpox in the principality of Serbia. Following a smallpox epidemic in 1837 that caused 40,000 deaths, the British government initiated a concentrated vaccination policy, starting with the Vaccination Act of 1840, which provided for universal vaccination and prohibited variolation. The Vaccination Act 1853 introduced compulsory smallpox vaccination in England and Wales. The law followed a severe outbreak of smallpox in 1851 and 1852. It provided that the poor law authorities would continue to dispense vaccination to all free of charge, but that records were to be kept on vaccinated children by the network of births registrars. It was accepted at the time, that voluntary vaccination had not reduced smallpox mortality, but the Vaccination Act 1853 was so badly implemented that it had little impact on the number of children vaccinated in England and Wales.

The U.S. Supreme Court upheld compulsory vaccination laws in the 1905 landmark case Jacobson v. Massachusetts, ruling that laws could require vaccination to protect the public from dangerous communicable diseases. However, in practice the U.S. had the lowest rate of vaccination among industrialized nations in the early 20th century. Compulsory vaccination laws began to be enforced in the U.S. after World War II. In 1959, the WHO called for the eradication of smallpox worldwide, as smallpox was still endemic in 33 countries. In the 1960s six to eight children died each year in the U.S. from vaccination-related complications. According to the WHO there were in 1966 about 100 million cases of smallpox worldwide, causing an estimated two million deaths. In the 1970s there was such a small risk of contracting smallpox that the U.S. Public Health Service recommended for routine smallpox vaccination to be ended. By 1974 the WHO smallpox vaccination program had confined smallpox to parts of Pakistan, India, Bangladesh, Ethiopia and Somalia. In 1977 the WHO recorded the last case of smallpox infection acquired outside a laboratory in Somalia. In 1980 the WHO officially declared the world free of smallpox.

In 1974 the WHO adopted the goal of universal vaccination by 1990 to protect children against six preventable infectious diseases: measles, poliomyelitis, diphtheria, whooping cough, tetanus, and tuberculosis. In the 1980s only 20 to 40% of children in developing countries were vaccinated against these six diseases. In wealthy nations the number of measles cases had dropped dramatically after the introduction of the measles vaccine in 1963. WHO figures demonstrate that in many countries a decline in measles vaccination leads to a resurgence in measles cases. Measles are so contagious that public health experts believe a vaccination rate of 100% is needed to control the disease. Despite decades of mass vaccination polio remains a threat in India, Nigeria, Somalia, Niger, Afghanistan, Bangladesh and Indonesia. By 2006 global health experts concluded that the eradication of polio was only possible if the supply of drinking water and sanitation facilities were improved in slums. The deployment of a combined DPT vaccine against diphtheria, pertussis (whooping cough), and tetanus in the 1950s was considered a major advancement for public health. But in the course of vaccination campaigns that spanned decades, DPT vaccines became associated with large number of cases with side effects. Despite improved DPT vaccines coming onto the market in the 1990s DPT vaccines became the focus of anti-vaccination campaigns in wealthy nations. As immunization rates fell outbreaks of pertussis increased in many countries.

In 2000, the Global Alliance for Vaccines and Immunization was established to strengthen routine vaccinations and introduce new and underused vaccines in countries with a per capita GDP of under US$1000.

UNICEF has reported on the extent to which children missed out on vaccinations from 2020 onwards due to the COVID-19 pandemic. By summer 2023, the organisation described vaccination programs as getting "back on track".

Mechanism of Function

Vaccines are a way of artificially activating the immune system to protect against infectious disease. The activation occurs through priming the immune system with an immunogen. Stimulating immune responses with an infectious agent is known as *immunization*. Vaccination includes various ways of administering immunogens.

Most vaccines are administered before a patient has contracted a disease to help increase future protection. However, some vaccines are administered after the patient already has contracted a disease. Vaccines given after exposure to smallpox are reported to offer some protection from disease or may reduce the severity of disease. The first rabies immunization was given by Louis Pasteur to a child after he was bitten by a rabid dog. Since its discovery, the rabies vaccine has been proven effective in preventing rabies in humans when administered several times over 14 days along with rabies immune globulin and wound care. Other examples include experimental AIDS, cancer and Alzheimer's disease vaccines. Such immunizations aim to trigger an immune response more rapidly and with less harm than natural infection.

Most vaccines are given by injection as they are not absorbed reliably through the intestines. Live attenuated polio, rotavirus, some typhoid, and some cholera vaccines are given orally to produce immunity in the bowel. While vaccination provides a lasting effect, it usually takes several weeks to develop. This differs from passive immunity (the transfer of antibodies, such as in breastfeeding), which has immediate effect.

A vaccine failure is when an organism contracts a disease in spite of being vaccinated against it. Primary vaccine failure occurs when an organism's immune system does not produce antibodies when first vaccinated. Vaccines can fail when several series are given and fail to produce an immune response. The term "vaccine failure" does not necessarily imply that the vaccine is defective. Most vaccine failures are simply due to individual variations in immune response.

Vaccination versus Inoculation

The term "inoculation" is often used interchangeably with "vaccination." However, while related, the terms are not synonymous. Vaccination is treatment of an individual with an attenuated (i.e. less virulent) pathogen or other immunogen, whereas inoculation, also called variolation in the context of smallpox prophylaxis, is treatment with unattenuated variola virus taken from a pustule or scab of a smallpox patient into the superficial layers of the skin, commonly the upper arm. Variolation was often done 'arm-to-arm' or, less effectively, 'scab-to-arm', and often caused the patient to become infected with smallpox, which in some cases resulted in severe disease. Vaccinations began in the late 18th century with the work of Edward Jenner and the smallpox vaccine.

Preventing Disease versus Preventing Infection

Some vaccines, like the smallpox vaccine, prevent infection. Their use results in sterilizing immunity and can help eradicate a disease if there is no animal reserve. Other vaccines, including those for COVID-19, help to (temporarily) lower the chance of severe disease for individuals, without necessarily reducing the probability of becoming infected.

Safety

Vaccine Development and Approval

Just like any medication or procedure, no vaccine can be 100% safe or effective for everyone because each person's body can react differently. While minor side effects, such as soreness or low grade fever, are relatively common, serious side effects are very rare and occur in about 1 out of every 100,000 vaccinations and typically involve allergic reactions that can cause hives or difficulty breathing.

However, vaccines are the safest they ever have been in history and each vaccine undergoes rigorous clinical trials to ensure their safety and efficacy before approval by authorities such as the US Food and Drug Administration (FDA).

Prior to human testing, vaccines are tested on cell cultures and the results modelled to assess how they will interact with the immune system. During the next round of testing, researchers study vaccines in animals, including mice, rabbits, guinea pigs, and monkeys. Vaccines that pass each of these stages of testing are then approved by the public health safety authority (FDA in the United States) to start a three-phase series of human testing, advancing to higher phases only if they are deemed safe and effective at the previous phase.

The people in these trials participate voluntarily and are required to prove they understand the purpose of the study and the potential risks.

During phase I trials, a vaccine is tested in a group of about 20 people with the primary goal of assessing the vaccine's safety. Phase II trials expand the testing to include 50 to several hundred people.

During this stage, the vaccine's safety continues to be evaluated and researchers also gather data on the effectiveness and the ideal dose of the vaccine. Vaccines determined to be safe and efficacious then advance to phase III trials, which focuses on the efficacy of the vaccine in hundreds to thousands of volunteers. This phase can take several years to complete and researchers use this opportunity to compare the vaccinated volunteers to those who have not been vaccinated to highlight any true reactions to the vaccine that occur.

If a vaccine passes all of the phases of testing, the manufacturer can then apply for license of the vaccine through the relevant regulatory authorities such as the FDA in US. Before regulatory authorities approve use in the general public, they extensively review the results of the clinical trials, safety tests, purity tests, and manufacturing methods and establish that the manufacturer itself is up to government standards in many other areas.

After regulatory approval, the regulators continue to monitor the manufacturing protocols, batch purity, and the manufacturing facility itself. Additionally, vaccines also undergo phase IV trials, which monitor the safety and efficacy of vaccines in tens of thousands of people, or more, across many years.

Side Effects

The Centers for Disease Control and Prevention (CDC) has compiled a list of vaccines and their possible side effects. The risk of side effects varies between vaccines.

Notable Vaccine Investigations

In 1976 in the United States, a mass swine flu vaccination programme was discontinued after 362 cases of Guillain–Barré syndrome among 45 million vaccinated people. William Foege of the CDC estimated that the incidence of Guillain-Barré was four times higher in vaccinated people than in those not receiving the swine flu vaccine.

Dengvaxia, the only approved vaccine for Dengue fever, was found to increase the risk of hospitalization for Dengue fever by 1.58 times in children of 9 years or younger, resulting in the suspension of a mass vaccination program in the Philippines in 2017.

Pandemrix – a vaccine for the H1N1 pandemic of 2009 given to around 31 million people – was found to have a higher level of adverse events than alternative vaccines resulting in legal action. In a response to the narcolepsy reports following immunization with Pandemrix, the CDC carried out a population-based study and found the FDA-approved 2009 H1N1 flu shots were not associated with an increased risk for the neurological disorder.

Monitoring

- CDC Immunization Safety Office initiatives.
- **Vaccine Adverse Event Reporting System (VAERS):** Food and Drug Administration (FDA) Center for Biologics Evaluation and Research (CBER) |Immunization Action Coalition (IAC).
- **Vaccine Safety Datalink (VSD):** Health Resources and Service Administration (HRSA) |Institute for Safe Medication Practices (ISMP).
- Clinical Immunization Safety Assessment (CISA) Project National Institutes of Health (NIH).
- National Vaccine Program Office (NVPO).
- The administration protocols, efficacy, and adverse events of vaccines are monitored by organizations of the US federal government, including the CDC and FDA, and independent agencies are constantly

re-evaluating vaccine practices. As with all medications, vaccine use is determined by public health research, surveillance, and reporting to governments and the public.

Usage

The World Health Organization (WHO) has estimated that vaccination prevents 3.5–5 million deaths per year, and up to 1.5 million children die each year due to diseases that could have been prevented by vaccination. They estimate that 29% of deaths of children under five-years-old in 2013 were vaccine preventable. In other developing parts of the world, they are faced with the challenge of having a decreased availability of resources and vaccinations. Countries such as those in Sub-Saharan Africa cannot afford to provide the full range of childhood vaccinations.

In 2024, a WHO/UNICEF report found "the number of children who received three doses of the vaccine against diphtheria, tetanus and pertussis (DTP) in 2023 – a key marker for global immunization coverage – stalled at 84% (108 million). However, the number of children who did not receive a single dose of the vaccine increased from 13.9 million in 2022 to 14.5 million in 2023. More than half of unvaccinated children live in the 31 countries with fragile, conflict-affected and vulnerable settings."

United States

Vaccines have led to major decreases in the prevalence of infectious diseases in the United States. In 2007, studies regarding the effectiveness of vaccines on mortality or morbidity rates of those exposed to various diseases have shown almost 100% decreases in death rates, and about a 90% decrease in exposure rates. This has allowed specific organizations and states to adopt standards for recommended early childhood vaccinations. Lower income families who are unable to otherwise afford vaccinations are supported by these organizations and specific government laws. The Vaccines for Children Program and the Social Security Act are two major players in supporting lower socioeconomic groups.

In 2000, the CDC declared that measles had been eliminated in the US (defined as no disease transmission for 12 continuous months). However, with the growing anti-vaccine movement, the US has seen a resurgence of certain vaccine-preventable diseases. The measles virus has now lost its elimination status in the US as the number of measles cases continues to rise in recent years with a total of 17 outbreaks in 2018 and 465 outbreaks in 2019 (as of 4 April 2019).

Vaccination Policy

To eliminate the risk of outbreaks of some diseases, at various times governments and other institutions have employed policies requiring vaccination for all people. For example, an 1853 law required universal vaccination against smallpox in

England and Wales, with fines levied on people who did not comply. Common contemporary U.S. vaccination policies require that children receive recommended vaccinations before entering public school.

Beginning with early vaccination in the nineteenth century, these policies were resisted by a variety of groups, collectively called antivaccinationists, who object on scientific, ethical, political, medical safety, religious, and other grounds. Common objections are that vaccinations do not work, that compulsory vaccination constitutes excessive government intervention in personal matters, or that the proposed vaccinations are not sufficiently safe. Many modern vaccination policies allow exemptions for people who have compromised immune systems, allergies to the components used in vaccinations or strongly held objections.

In countries with limited financial resources, limited vaccination coverage results in greater morbidity and mortality due to infectious disease. More affluent countries are able to subsidize vaccinations for at-risk groups, resulting in more comprehensive and effective coverage. In Australia, for example, the Government subsidizes vaccinations for seniors and indigenous Australians.

Public Health Law Research, an independent US based organization, reported in 2009 that there is insufficient evidence to assess the effectiveness of requiring vaccinations as a condition for specified jobs as a means of reducing incidence of specific diseases among particularly vulnerable populations; that there is sufficient evidence supporting the effectiveness of requiring vaccinations as a condition for attending child care facilities and schools; and that there is strong evidence supporting the effectiveness of standing orders, which allow healthcare workers without prescription authority to administer vaccine as a public health intervention.

Fractional Dose Vaccination

Fractional dose vaccination reduces the dose of a vaccine to allow more individuals to be vaccinated with a given vaccine stock, trading societal benefit for individual protection. Based on the nonlinearity properties of many vaccines, it is effective in poverty diseases and promises benefits in pandemic waves, e.g. in COVID-19, when vaccine supply is limited.

Litigation

Allegations of vaccine injuries in recent decades have appeared in litigation in the U.S. Some families have won substantial awards from sympathetic juries, even though most public health officials have said that the claims of injuries were unfounded. In response, several vaccine makers stopped production, which the US government believed could be a threat to public health, so laws were passed to shield manufacturers from liabilities stemming from vaccine injury claims.

The safety and side effects of multiple vaccines have been tested to uphold the viability of vaccines as a barrier against disease. The influenza vaccine was tested in controlled trials and proven to have negligible side effects equal to that of a placebo. Some concerns from families might have arisen from social beliefs and norms that cause them to mistrust or refuse vaccinations, contributing to this discrepancy in side effects that were unfounded.

Opposition

Opposition to vaccination, from a wide array of vaccine critics, has existed since the earliest vaccination campaigns. It is widely accepted that the benefits of preventing serious illness and death from infectious diseases greatly outweigh the risks of rare serious adverse effects following immunization. Some studies have claimed to show that current vaccine schedules increase infant mortality and hospitalization rates; those studies, however, are correlational in nature and therefore cannot demonstrate causal effects, and the studies have also been criticized for cherry picking the comparisons they report, for ignoring historical trends that support an opposing conclusion, and for counting vaccines in a manner that is "completely arbitrary and riddled with mistakes".

Various disputes have arisen over the morality, ethics, effectiveness, and safety of vaccination. Some vaccination critics say that vaccines are ineffective against disease or that vaccine safety studies are inadequate. Some religious groups do not allow vaccination, and some political groups oppose mandatory vaccination on the grounds of individual liberty.

In response, concern has been raised that spreading unfounded information about the medical risks of vaccines increases rates of life-threatening infections, not only in the children whose parents refused vaccinations, but also in those who cannot be vaccinated due to age or immunodeficiency, who could contract infections from unvaccinated carriers. Some parents believe vaccinations cause autism, although there is no scientific evidence to support this idea. In 2011, Andrew Wakefield, a leading proponent of the theory that MMR vaccine causes autism, was found to have been financially motivated to falsify research data and was subsequently stripped of his medical license. In the United States people who refuse vaccines for non-medical reasons have made up a large percentage of the cases of measles, and subsequent cases of permanent hearing loss and death caused by the disease.

Many parents do not vaccinate their children because they feel that diseases are no longer present due to vaccination. This is a false assumption, since diseases held in check by immunization programs can and do still return if immunization is dropped. These pathogens could possibly infect vaccinated people, due to the pathogen's ability to mutate when it is able to live in unvaccinated hosts.

Vaccination and Autism

The notion of a connection between vaccines and autism originated in a 1998 paper published in *The Lancet* whose lead author was the physician Andrew Wakefield. His study concluded that eight of the 12 patients, ages 3 years to 10 years, developed behavioral symptoms consistent with autism following the MMR vaccine (an immunization against measles, mumps, and rubella). The article was widely criticized for lack of scientific rigor and it was proven that Wakefield falsified data in the article. In 2004, 10 of the original 12 co-authors (not including Wakefield) published a retraction of the article and stated the following: "We wish to make it clear that in this paper no causal link was established between MMR vaccine and autism as the data were insufficient." In 2010, *The Lancet* officially retracted the article, stating that several elements of the article were incorrect, including falsified data and protocols. The article has sparked a much greater anti-vaccination movement, particularly in the United States, and even though the article was shown to be fraudulent and was heavily retracted, one in four parents still believe that vaccines can cause autism.

To date, all validated and definitive studies have shown that there is no correlation between vaccines and autism. One of the studies published in 2015 confirms there is no link between autism and the MMR vaccine. Infants were given a health plan, that included an MMR vaccine, and were continuously studied until they reached five years old. There was no link between the vaccine and children who had a normally developed sibling or a sibling that had autism making them a higher risk for developing autism themselves.

It can be difficult to correct the memory of humans when wrong information is received prior to correct information. Even though there is much evidence to go against the Wakefield study and retractions were published by most of the co-authors, many people continue to believe and base decisions on the study as it still lingers in their memory. Studies and research are being conducted to determine effective ways to correct misinformation in the public memory.

Routes of Administration

A vaccine administration may be oral, by injection (intramuscular, intradermal, subcutaneous), by puncture, transdermal or intranasal. Several recent clinical trials have aimed to deliver the vaccines via mucosal surfaces to be up-taken by the common mucosal immunity system, thus avoiding the need for injections.

Economics of Vaccination

Health is often used as one of the metrics for determining the economic prosperity of a country. This is because healthier individuals are generally better suited to contributing to the economic development of a country than the sick. There are many reasons for this. For instance, a person who is vaccinated for influenza not only protects themselves from the risk of influenza, but simultaneously also prevents themselves from infecting those around them. This leads to a healthier society, which allows individuals to be more economically productive. Children are consequently able to attend school more often and have been shown to do better academically. Similarly, adults are able to work more often, more efficiently, and more effectively.

Costs and Benefits

On the whole, vaccinations induce a net benefit to society. Vaccines are often noted for their high Return on investment (ROI) values, especially when considering the long-term effects. Some vaccines have much higher ROI values than others. Studies have shown that the ratios of vaccination benefits to costs can differ substantially—from 27:1 for diphtheria/pertussis, to 13.5:1 for measles, 4.76:1 for varicella, and 0.68–1.1 : 1 for pneumococcal conjugate. Some governments choose to subsidize the costs of vaccines, due to some of the high ROI values attributed to vaccinations. The United States subsidizes over half of all vaccines for children, which costs between $400 and $600 each. Although most children do get vaccinated, the adult population of the US is still below the recommended immunization levels. Many factors can be attributed to this issue. Many adults who have other health conditions are unable to be safely immunized, whereas others opt not to be immunized for the sake of private financial benefits. Many Americans are underinsured, and, as such, are required to pay for vaccines out-of-pocket. Others are responsible for paying high deductibles and co-pays. Although vaccinations usually induce long-term economic benefits, many governments struggle to pay the high short-term costs associated with labor and production. Consequently, many countries neglect to provide such services.

According to a 2021 paper, vaccinations against haemophilus influenzae type b, hepatitis B, human papillomavirus, Japanese encephalitis, measles, neisseria meningitidis serogroup A, rotavirus, rubella, streptococcus pneumoniae, and yellow fever have prevented an estimated 50 million deaths from 2000 to 2019. The paper "represents the largest assessment of vaccine impact before COVID-19-related disruptions". According to a June 2022 study, COVID19 vaccinations prevented an additional 14.4 to 19.8 million deaths in 185 countries and territories from 8 December 2020 to 8 December 2021.

They estimated that it would cost between $2.8 billion and $3.7 billion to develop at least one vaccine for each of them. This should be set against the potential cost of an outbreak. The 2003 SARS outbreak in East Asia cost $54 billion.

Game theory uses utility functions to model costs and benefits, which may include financial and non-financial costs and benefits. In recent years, it has been argued that game theory can effectively be used to model vaccine uptake in societies. Researchers have used game theory for this purpose to analyse vaccination uptake in the context of diseases such as influenza and measles.

10.1.2 Immunization

Immunization, or immunisation, is the process by which an individual's immune system becomes fortified against an infectious agent (known as the immunogen). When this system is exposed to molecules that are foreign to the body, called *non-self*, it will orchestrate an immune response, and it will also develop the ability to quickly respond to a subsequent encounter because of immunological memory. This is a function of the adaptive immune system. Therefore, by exposing a human, or an animal, to an immunogen in a controlled way, its body can learn to protect itself: this is called active immunization.

The most important elements of the immune system that are improved by immunization are the T cells, B cells, and the antibodies B cells produce. Memory B cells and memory T cells are responsible for a swift response to a second encounter with a foreign molecule. Passive immunization is direct introduction of these elements into the body, instead of production of these elements by the body itself.

Immunization happens in various ways, both in the wild and as done by human efforts in health care. Natural immunity is gained by those organisms whose immune systems succeed in fighting off a previous infection, if the relevant pathogen is one for which immunization is even possible. Natural immunity can have degrees of effectiveness (partial rather than absolute) and may fade over time (within months, years, or decades, depending on the pathogen). In health care, the main technique of artificial induction of immunity is vaccination, which is a major form of prevention of disease, whether by prevention of infection (pathogen fails to mount sufficient reproduction in the host), prevention of severe disease (infection still happens but is not severe), or both. Vaccination against vaccine-preventable diseases is a major relief of disease burden even though it usually cannot eradicate a disease. Vaccines against microorganisms that cause diseases can prepare the body's immune system, thus helping to fight or prevent an infection. The fact that mutations can cause cancer cells to produce proteins or other molecules that are known to the body forms the theoretical basis for therapeutic cancer vaccines. Other molecules can be used for immunization as well, for example in experimental vaccines against nicotine (NicVAX) or the hormone ghrelin in experiments to create an obesity vaccine.

Immunizations are often widely stated as less risky and an easier way to become immune to a particular disease than risking a milder form of the disease itself. They are important for both adults and children in that they can protect us from the many diseases out there. Immunization not only protects children against deadly diseases but also helps in developing children's immune systems. Through the use of immunizations, some infections and diseases have almost completely been eradicated throughout the World. One example is polio. Thanks to dedicated health care professionals and the parents of children who vaccinated on schedule, polio has been eliminated in the U.S., since 1979. Polio is still found in other parts of the world so certain people could still be at risk of getting it. This includes those people who have never had the vaccine, those who did not receive all doses of the vaccine, or those traveling to areas of the world where polio is still prevalent. Active immunization/vaccination has been named one of the "Ten Great Public Health Achievements in the 20th Century".

History

Before the introduction of vaccines, people could only become immune to an infectious disease by contracting the disease and surviving it. Smallpox (variola) was prevented in this way by inoculation, which produced a milder effect than the natural disease. The first clear reference to smallpox inoculation was made by the Chinese author Wan Quan (1499–1582) in his *Douzhen xinfa* published in 1549. In China, powdered smallpox scabs were blown up the noses of the healthy. The patients would then develop a mild case of the disease and from then on were immune to it. The technique did have a 0.5–2.0% mortality rate, but that was considerably less than the 20–30% mortality rate of the disease itself. Two reports on the Chinese practice of inoculation were received by the Royal Society in London in 1700; one by Dr. Martin Lister who received a report by an employee of the East India Company stationed in China and another by Clopton Havers. According to Voltaire (1742), the Turks derived their use of inoculation from neighbouring Circassia. Voltaire does not speculate on where the Circassians derived their technique from, though he reports that the Chinese have practiced it "these hundred years". It was introduced into England from Turkey by Lady Mary Wortley Montagu in 1721 and used by Zabdiel Boylston in Boston the same year. In 1798 Edward Jenner introduced inoculation with cowpox (smallpox vaccine), a much safer procedure. This procedure, referred to as vaccination, gradually replaced smallpox inoculation, now called variolation to distinguish it from vaccination. Until the 1880s vaccine/vaccination referred only to smallpox, but Louis Pasteur developed immunization methods for chicken cholera and anthrax in animals and for human rabies, and suggested that the terms vaccine/vaccination should be extended to cover the new procedures. This can cause confusion if care is not taken to specify which vaccine is used e.g. measles vaccine or influenza vaccine.

Active Immunization

Active immunization can occur naturally when a person comes in contact with, for example, a microbe. The immune system will eventually create antibodies and other defenses against the microbe. The next time, the immune response against this microbe can be very efficient; this is the case in many of the childhood infections that a person only contracts once, but then is immune.

Artificial active immunization is where the microbe, or parts of it, are injected into the person before they are able to take it in naturally. If whole microbes are used, they are pre-treated.

The importance of immunization is so great that the American Centers for Disease Control and Prevention has named it one of the "Ten Great Public Health Achievements in the 20th Century". Live attenuated vaccines have decreased pathogenicity. Their effectiveness depends on the immune systems ability to replicate and elicits a response similar to natural infection. It is usually effective with a single dose. Examples of live, attenuated vaccines include measles, mumps, rubella, MMR, yellow fever, varicella, rotavirus, and influenza (LAIV).

Passive Immunization

Passive immunization is where pre-synthesized elements of the immune system are transferred to a person so that the body does not need to produce these elements itself. Currently, antibodies can be used for passive immunization. This method of immunization begins to work very quickly, but it is short lasting, because the antibodies are naturally broken down, and if there are no B cells to produce more antibodies, they will disappear.

Passive immunization occurs physiologically, when antibodies are transferred from mother to fetus during pregnancy, to protect the fetus before and shortly after birth.

Artificial passive immunization is normally administered by injection and is used if there has been a recent outbreak of a particular disease or as an emergency treatment for toxicity, as in for tetanus. The antibodies can be produced in animals, called "serum therapy," although there is a high chance of anaphylactic shock because of immunity against animal serum itself. Thus, humanized antibodies produced *in vitro* by cell culture are used instead if available.

Economics of Immunizations

Positive Externality

Immunizations impose what is known as a positive consumer externality on society. In addition to providing the individual with protection against certain antigens it adds greater protection to all other individuals in society through herd immunity. Because this extra protection is not accounted for in the market transactions for immunizations we see an undervaluing of the marginal benefit of each immunization. This market failure is caused by individuals making decisions based on their private marginal benefit instead of the social marginal benefit. Society's undervaluing of immunizations means that through normal market transactions we end up at a quantity that is lower than what is socially optimal.

For example, if individual A values their own immunity to an antigen at $100 but the immunization costs $150, individual A will decide against receiving immunization. However, if the added benefit of herd immunity means person B values person A's immunity at $70 then the total social marginal benefit of their immunization is $170. Individual A's private marginal benefit being lower than the social marginal benefit leads to an under-consumption of immunizations.

Socially Optimal Outcome

Having private marginal benefits lower than social marginal benefits will always lead to an under-consumption of any good. The size of the disparity is determined by the value that society places on each different immunization. Many times, immunizations do not reach a socially optimum quantity high enough to eradicate the antigen. Instead, they reach a social quantity that allows for an optimal amount of sick individuals. Most of the commonly immunized diseases in the United States still see a small presence with occasional larger outbreaks. Measles is a good example of a disease whose social optimum leaves enough room for outbreaks in the United States that often lead to the deaths of a handful of individuals.

There are also examples of illnesses so dangerous that the social optimum ended with the eradication of the virus, such as smallpox. In these cases, the social marginal benefit is so large that society is willing to pay the cost to reach a level of immunization that makes the spread and survival of the disease impossible.

Despite the severity of certain illnesses, the cost of immunization versus the social marginal benefit means that total eradication is not always the end goal of immunization. Though it is hard to tell exactly where the socially optimal outcome is, we know that it is not the eradication of all disease for which an immunization exists.

Internalizing the Externality

In order to internalize the positive externality imposed by immunizations payments equal to the marginal benefit must be made. In countries like the United States these payment usually come in the form of subsidies from the government. Before 1962 immunization programs in the United States were run on the local and state level of governments. The inconsistency in subsidies lead to some regions of the United States reaching the socially optimal quantity while other regions were left without subsidies and remained at the private marginal benefit level of immunizations. Since 1962 and the Vaccination Assistance Act, the United States as a whole has been moving towards the socially optimal outcome on a larger scale.

Despite government subsidies it is difficult to tell when social optimum has been achieved. In addition to hardships determining the true social marginal benefit of immunizations we see cultural movements shifting private marginal benefit curves. Vaccine controversies have changed the way some private citizens view the marginal benefit of being immunized. If Individual A believes that there is a large health risk, possibly larger than the antigen itself, associated with immunization they will not be willing to pay for or receive immunization. With fewer willing participants and a widening marginal benefit reaching a social optimum becomes more difficult for governments to achieve through subsidies.

Outside of government intervention through subsidies, non profit organizations can also move a society towards the socially optimal outcome by providing free immunizations to developing regions. Without the ability to afford the immunizations to begin with, developing societies will not be able to reach a quantity determined by private marginal benefits. By running immunization programs organizations are able to move privately under-immunized communities towards the social optimum.

Race, Ethnicity and Immunization

In the United States, race and ethnicity are strong determinants of utilization of preventive and therapeutic health services as well as health outcomes. Rates of infant mortality and most of the leading causes of overall mortality have been higher in African Americans than in European Americans. A recent analysis of mortality from influenza and pneumonia revealed that African Americans died of these causes at higher rates than European Americans in 1999–2018. Contributing to these racial disparities are lower rates of immunization against influenza and pneumococcal pneumonia. During the COVID-19 pandemic, death rates have been higher in African Americans than European Americans and vaccination rates have lagged in African Americans during the roll-out. Among Hispanics immunization rates are lower than those in non-Hispanic whites.

10.2 THE SCIENCE BEHIND VACCINES

Vaccines work by training the immune system to recognize and combat specific pathogens. When a vaccine is administered, it introduces antigens into the body, which are harmless parts of the pathogen (such as proteins or inactivated viruses). These antigens stimulate the immune system to produce a response.

The immune system, in turn, produces **antibodies**—specialized proteins that can neutralize or mark pathogens for destruction by other immune cells. Additionally, **memory cells** (both B cells and T cells) are formed, allowing the immune system to remember the pathogen and mount a more rapid and efficient response if the body encounters the pathogen in the future.

There are different types of vaccines, including:

- **Inactivated (Killed) Vaccines:** These vaccines contain viruses or bacteria that have been killed or inactivated so they cannot cause disease. Examples include the polio vaccine (IPV) and the hepatitis A vaccine.
- **Live Attenuated Vaccines:** These vaccines contain weakened forms of the virus or bacteria that are still capable of replicating but are unable to cause disease in healthy individuals. Examples include the measles, mumps, and rubella (MMR) vaccine and the yellow fever vaccine.
- **Subunit, Recombinant, or Conjugate Vaccines:** These vaccines include only parts of the pathogen, such as a protein or sugar molecule, that is enough to elicit an immune response. Examples include the human papillomavirus (HPV) vaccine and the Haemophilus influenzae type b (Hib) vaccine.
- **mRNA Vaccines:** A newer approach, mRNA vaccines use messenger RNA to instruct cells to produce a protein from the pathogen, triggering an immune response. The Pfizer-BioNTech and Moderna COVID-19 vaccines are examples of this type.

10.3 BENEFITS OF VACCINATION AND IMMUNIZATION

Vaccination and immunization have revolutionized medicine and public health in several profound ways. Some of the key benefits include:

- **Prevention of Disease**: Vaccines provide immunity to many serious diseases, such as measles, smallpox, polio, diphtheria, and influenza. By preventing these diseases, vaccines have saved millions of lives and reduced morbidity worldwide.
- **Herd Immunity**: When a significant portion of a population is vaccinated, it can lead to herd immunity, where even those who are not vaccinated (e.g., infants, elderly, or immunocompromised individuals) are less likely to get infected because the spread of the disease is minimized. This protects the most vulnerable members of society.
- **Eradication of Diseases**: Through global vaccination campaigns, certain diseases, such as smallpox, have been completely eradicated, and others, like polio, are close to being eliminated. This achievement has had profound effects on reducing disease burden and healthcare costs.
- **Protection Against Emerging Diseases**: Vaccines also provide protection against new or emerging infectious diseases. The rapid development and distribution of the COVID-19 vaccines are an example

of how immunization can help control the spread of newly identified diseases.

- **Cost-Effectiveness:** Vaccines are often far more cost-effective than treating the diseases they prevent. By preventing illness, hospitalization, and long-term healthcare costs, vaccination programs reduce the economic burden on individuals, families, and healthcare systems.

10.4 CHALLENGES IN VACCINATION AND IMMUNIZATION

While vaccination has been immensely successful, there are still several challenges that need to be addressed to achieve global immunization goals:

- **Vaccine Hesitancy:** One of the significant challenges in vaccination is vaccine hesitancy, which is the reluctance or refusal to vaccinate despite availability of vaccines. This can result from misinformation, fear of side effects, religious or cultural beliefs, or distrust in the medical system. Vaccine hesitancy has led to outbreaks of preventable diseases, such as measles, in communities that were previously disease-free.
- **Access to Vaccines:** While vaccines are widely available in many developed countries, access to immunization remains a major challenge in low-income and developing regions. Factors such as lack of healthcare infrastructure, political instability, and inadequate distribution networks hinder vaccination efforts. The Global Alliance for Vaccines and Immunization (GAVI) has been working to address these disparities by improving vaccine access in the world's poorest countries.
- **Vaccine Production and Supply:** The production and distribution of vaccines are resource-intensive, and in times of global health crises, such as the COVID-19 pandemic, ensuring a sufficient supply of vaccines for all countries can be difficult. This includes not only vaccine production but also cold chain management and logistics.
- **Side Effects and Safety:** Although vaccines are rigorously tested for safety and efficacy, like any medical intervention, they may cause side effects in some individuals. However, the benefits of vaccination far outweigh the risks. Public education is essential to communicate the safety of vaccines and address concerns.

10.5 THE ROLE OF VACCINATION IN DISEASE CONTROL

Vaccination has played a critical role in controlling infectious diseases worldwide. Some vaccines, such as those for smallpox and polio, have led to complete eradication or near-eradication of these diseases, demonstrating the immense potential of immunization to change the course of human history.

- **Global Immunization Programs**: International efforts, such as the Expanded Programme on Immunization (EPI) led by the World Health Organization (WHO), have contributed to the widespread delivery of vaccines, particularly in low- and middle-income countries. These programs aim to ensure that every child receives the vaccines needed to protect against common childhood diseases, such as measles, diphtheria, and tetanus.
- **Vaccination During Emergencies**: Vaccines play a crucial role in emergency response during disease outbreaks. For instance, in the wake of the Ebola outbreak in West Africa, the rapid development and deployment of the Ebola vaccine helped control the spread of the virus. Similarly, the development of COVID-19 vaccines was a monumental achievement in controlling a global pandemic.

10.6 THE FUTURE OF VACCINATION AND IMMUNIZATION

The future of vaccination and immunization holds great promise. Advancements in vaccine technology, such as mRNA vaccines and nanoparticle-based vaccines, are improving the speed and efficiency with which vaccines can be developed, tested, and distributed. These innovations will likely be crucial in addressing future public health challenges, including the prevention of emerging infectious diseases and tackling diseases for which vaccines are not yet available. Additionally, global efforts to increase vaccine coverage, reduce vaccine hesitancy, and ensure equitable access to vaccines will continue to be essential in improving public health worldwide.

10.7 CONCLUSION

Vaccination and immunization are among the most effective tools in preventing infectious diseases and improving public health. Their benefits, including disease prevention, herd immunity, and eradication of diseases, are immeasurable. However, challenges such as vaccine hesitancy, access to vaccines, and production capacity must be addressed to ensure that all people, regardless of location or economic status, can benefit from the protection that vaccines offer. As science continues to advance, vaccines will play an even greater role in global health, potentially eradicating more diseases and saving millions of lives.

PART 4:

PRESCRIBING THERAPEUTICS AND INFECTION CONTROL

CHAPTER 11

ANTIBIOTICS AND SYNTHETIC ANTIMICROBIAL AGENTS: THEIR PROPERTIES AND USES

Antibiotics and synthetic antimicrobial agents have been transformative in the field of medicine, significantly reducing the mortality rates associated with bacterial infections and revolutionizing treatment protocols. Their discovery and development have paved the way for modern medicine, but as bacteria evolve and resistance increases, understanding their properties and appropriate uses has become more crucial than ever.

11.1 UNDERSTANDING ANTIBIOTICS AND ANTIMICROBIAL AGENTS

Antibiotics are naturally occurring substances or compounds derived from microorganisms that can inhibit the growth of or kill bacteria. They have been pivotal in treating infections caused by bacterial pathogens. Antibiotics, initially sourced from molds or bacteria, like *Penicillium* and *Streptomyces*, were a breakthrough discovery in the early 20th century. **Synthetic antimicrobial agents**, on the other hand, are chemically engineered substances designed to inhibit or kill microorganisms, including bacteria, fungi, and viruses. While they have antimicrobial activity, synthetic agents are often broader in their scope than natural antibiotics, targeting a range of pathogens, including those that antibiotics may not be effective against. These agents are typically created in laboratories through chemical synthesis and often serve as alternatives to or complements for naturally occurring antibiotics.

11.2 PROPERTIES OF ANTIBIOTICS AND ANTIMICROBIAL AGENTS

The effectiveness of antibiotics and antimicrobial agents is determined by several properties, which include their spectrum of activity, pharmacokinetics, mode of action, and toxicity.

- **Spectrum of Activity**: This refers to the range of microorganisms against which an antimicrobial agent is effective. Antibiotics can be classified as **broad-spectrum** or **narrow-spectrum**:

- *Broad-Spectrum Antibiotics:* These affect a wide variety of bacteria, both Gram-positive and Gram-negative. Examples include tetracyclines and ciprofloxacin. They are useful in treating infections when the specific causative organism is unknown, but their indiscriminate use can lead to increased risk of resistance.
- *Narrow-Spectrum Antibiotics:* These specifically target certain bacteria or groups of bacteria. Penicillin, for example, is typically effective against Gram-positive bacteria. Narrow-spectrum antibiotics are preferable when the pathogen is known, as they minimize damage to the normal microbiota.

- **Mode of Action:** Antibiotics and synthetic agents exert their effect by targeting specific structures or processes within bacterial cells:
  - *Inhibition of Cell Wall Synthesis:* Penicillin and cephalosporins work by preventing bacterial cell wall formation, leading to cell lysis.
  - *Protein Synthesis Inhibition:* Tetracyclines and macrolides block the synthesis of proteins in bacteria, preventing their growth and replication.
  - *DNA Replication and Transcription Inhibition:* Fluoroquinolones, like ciprofloxacin, target bacterial enzymes involved in DNA replication, hindering bacterial reproduction.
  - *Disruption of Cell Membrane Integrity:* Polymyxins disrupt the bacterial cell membrane, leading to leakage of cellular contents.
- **Pharmacokinetics:** This encompasses how the body absorbs, distributes, metabolizes, and excretes the antibiotic. It determines the drug's effectiveness and appropriate dosing. Some antibiotics require administration multiple times a day, while others maintain effective levels in the body for extended periods.
- **Toxicity:** Antibiotics and synthetic agents can have side effects on the host. These can range from mild allergic reactions to more severe effects, such as liver damage or kidney toxicity, depending on the drug and its usage.

11.3 CLASSIFICATION OF ANTIBIOTICS AND SYNTHETIC ANTIMICROBIAL AGENTS

Antibiotics and synthetic agents can be classified based on their chemical structure, spectrum of activity, and mechanism of action. Common classes include:

- **Penicillins and Cephalosporins:** These are beta-lactam antibiotics that inhibit cell wall synthesis in bacteria. Penicillin was the first antibiotic discovered and has been widely used since the 1940s. Cephalosporins, which are structurally similar, are often used for a broader range of bacterial infections.

- **Aminoglycosides:** These antibiotics, including streptomycin and gentamicin, target bacterial ribosomes, inhibiting protein synthesis. They are used primarily for serious Gram-negative infections.
- **Tetracyclines:** These antibiotics inhibit protein synthesis by binding to bacterial ribosomes. Tetracyclines are broad-spectrum and are used to treat conditions like acne, respiratory infections, and Lyme disease.
- **Macrolides:** Such as erythromycin and azithromycin, macrolides are effective against a wide range of bacteria by inhibiting protein synthesis. They are often used for respiratory tract infections, skin infections, and sexually transmitted diseases.
- **Fluoroquinolones:** Fluoroquinolones, like ciprofloxacin, interfere with DNA replication and are broad-spectrum agents, particularly effective against Gram-negative bacteria.
- **Sulfonamides:** These are synthetic antimicrobial agents that interfere with bacterial folic acid synthesis. While not technically antibiotics (since they are chemically synthesized), they have similar antimicrobial properties.
- **Polymyxins:** These agents disrupt bacterial cell membranes and are typically used to treat multidrug-resistant Gram-negative infections.

11.4 USES OF ANTIBIOTICS AND SYNTHETIC ANTIMICROBIAL AGENTS

Antibiotics and synthetic antimicrobial agents are used to treat a wide range of infections caused by bacteria, fungi, and parasites. Some of their primary uses include:

- **Treatment of Bacterial Infections:** Antibiotics like penicillin and cephalosporins are commonly prescribed to treat bacterial infections such as pneumonia, urinary tract infections, and strep throat.
- **Prophylaxis:** Antibiotics are sometimes used preventively to avoid infections, particularly after surgeries, in immunocompromised patients, or in individuals exposed to particular infectious agents.
- **Management of Chronic Diseases:** In conditions such as tuberculosis, where prolonged antibiotic treatment is necessary, a combination of antibiotics may be used to ensure complete eradication of the pathogen.
- **Topical Antimicrobial Agents:** Synthetic antimicrobial agents, including mupirocin and silver sulfadiazine, are applied to the skin to prevent or treat localized infections, especially in burns or wounds.

- **Veterinary Medicine**: Antibiotics and synthetic agents are also widely used in veterinary medicine to treat animal infections and to prevent disease outbreaks in livestock.

11.5 ANTIMICROBIAL RESISTANCE: A GROWING CONCERN

One of the most pressing issues in modern medicine is the growing problem of **antimicrobial resistance (AMR)**. When bacteria evolve mechanisms to resist the effects of antibiotics, it leads to the emergence of **multidrug-resistant (MDR)** and **extensively drug-resistant (XDR)** strains, making infections harder to treat. Overuse and misuse of antibiotics, especially in humans and animals, have contributed significantly to the rise in resistance.

To combat AMR, it is essential to promote proper antibiotic stewardship, limit unnecessary prescriptions, and invest in research to develop new antibiotics and alternative treatments. Additionally, global cooperation is required to address resistance and promote the responsible use of antibiotics worldwide.

11.6 FUTURE DIRECTIONS IN ANTIBIOTIC AND ANTIMICROBIAL DEVELOPMENT

As bacteria continue to evolve and resistance spreads, the development of new antibiotics and synthetic antimicrobial agents remains a priority for researchers and pharmaceutical companies. Some promising areas of research include:

- **Antibiotics with Novel Mechanisms of Action**: Researchers are exploring new classes of antibiotics that work in ways that are fundamentally different from existing drugs. This includes agents that target bacterial virulence factors or bacterial communication systems, such as quorum sensing.
- **Bacteriophage Therapy**: Bacteriophages—viruses that infect and kill bacteria—are being explored as an alternative to antibiotics, particularly for treating multidrug-resistant infections.
- **Combination Therapies**: Combining existing antibiotics or using antibiotics in conjunction with adjuvants (substances that enhance the effect of the antibiotic) may help overcome resistance and improve treatment outcomes.
- **Nanotechnology**: Nanoparticles and nanomaterials are being studied for their ability to disrupt bacterial cell membranes, deliver drugs more effectively, and reduce the likelihood of resistance development.

11.7 CONCLUSION

Antibiotics and synthetic antimicrobial agents are indispensable tools in the fight against infectious diseases, offering life-saving treatments and prevention for bacterial, fungal, and parasitic infections.

Despite the critical role they play, the rise of antimicrobial resistance presents a formidable challenge that requires coordinated efforts to ensure their continued effectiveness.

Responsible usage, research into new agents, and the development of alternative treatments are essential steps in managing and overcoming this global health crisis. As the science of antimicrobial agents continues to evolve, new innovations hold the potential to combat the growing threat of resistant infections and improve global health outcomes.

CHAPTER 12

MECHANISMS OF ACTION OF ANTIBIOTICS AND SYNTHETIC ANTI-INFECTIVE AGENTS

Antibiotics and synthetic anti-infective agents have revolutionized the treatment of infectious diseases by inhibiting or killing pathogens. These agents target specific bacterial processes, making them effective against infections. Understanding the mechanisms by which antibiotics and synthetic anti-infective agents work is essential for their optimal use in clinical settings and for combating emerging challenges, such as antibiotic resistance.

12.1 INHIBITION OF CELL WALL SYNTHESIS

The bacterial cell wall provides structural integrity and protection against osmotic pressure. In many bacteria, the cell wall is made of peptidoglycan, a polymer consisting of sugars and amino acids. Several antibiotics target the synthesis of this peptidoglycan layer, leading to cell lysis and death.

- **Beta-lactam Antibiotics:** Penicillins, cephalosporins, and other beta-lactam antibiotics are among the most widely used antibiotics. They work by inhibiting enzymes known as **penicillin-binding proteins (PBPs)**, which are crucial for the final steps of peptidoglycan synthesis. The inhibition of these enzymes prevents the proper formation of the bacterial cell wall, resulting in bacterial cell death due to osmotic lysis.
- **Glycopeptides:** Antibiotics like vancomycin and teicoplanin inhibit cell wall synthesis by binding to the precursor molecules involved in peptidoglycan production. This prevents the polymerization of the peptidoglycan, leading to weakened cell walls and cell death. These antibiotics are particularly effective against Gram-positive bacteria.

12.2 INHIBITION OF PROTEIN SYNTHESIS

Protein synthesis is a vital process in bacterial growth and replication, and many antibiotics interfere with this function by targeting bacterial ribosomes. Ribosomes are the molecular machines responsible for translating mRNA into proteins. Antibiotics that inhibit protein synthesis bind to the bacterial ribosome and either block or disrupt the process of translation.

- **Aminoglycosides:** Examples include gentamicin and streptomycin. Aminoglycosides bind to the 30S subunit of the bacterial ribosome, causing misreading of the mRNA, leading to the production of faulty proteins. This results in bactericidal effects. Aminoglycosides are primarily effective against Gram-negative bacteria.
- **Tetracyclines:** Tetracyclines, such as doxycycline, bind to the 30S ribosomal subunit and block the attachment of tRNA to the ribosome. This inhibition prevents the elongation of the polypeptide chain during protein synthesis, which disrupts bacterial growth. Tetracyclines have broad-spectrum activity and are used to treat various bacterial infections.
- **Macrolides:** Antibiotics such as erythromycin and azithromycin bind to the 50S subunit of the bacterial ribosome, inhibiting the elongation step of protein synthesis. This prevents the ribosome from moving along the mRNA and producing proteins. Macrolides are effective against a variety of Gram-positive and some Gram-negative bacteria.
- **Chloramphenicol:** This antibiotic binds to the 50S ribosomal subunit and inhibits peptidyl transferase activity, preventing the addition of new amino acids to the growing protein chain. While effective against a broad range of bacteria, its use is limited due to potential side effects, such as aplastic anemia.

12.3 DISRUPTION OF CELL MEMBRANE INTEGRITY

The bacterial cell membrane is essential for maintaining the integrity of the cell, controlling the transport of nutrients, and maintaining cellular homeostasis. Disruption of the cell membrane leads to leakage of cellular contents and bacterial death.

- **Polymyxins:** Polymyxin B and polymyxin E (colistin) are cyclic peptides that bind to the outer membrane of Gram-negative bacteria. They interact with the phospholipids in the membrane, disrupting its integrity and causing leakage of cell contents. These antibiotics are effective against multidrug-resistant Gram-negative bacteria but are typically used as a last resort due to their nephrotoxicity.
- **Lipopeptides:** Daptomycin is a lipopeptide antibiotic that inserts into the bacterial cell membrane and forms pores, leading to a loss of membrane potential. This results in the disruption of bacterial cell function and eventual cell death. Daptomycin is used to treat infections caused by Gram-positive bacteria, including methicillin-resistant *Staphylococcus aureus* (MRSA).

12.4 INHIBITION OF NUCLEIC ACID SYNTHESIS

Bacterial DNA replication, transcription, and repair are essential processes for bacterial survival and multiplication. Several antibiotics interfere with these processes by targeting enzymes involved in nucleic acid synthesis.

- **Quinolones and Fluoroquinolones:** Ciprofloxacin and levofloxacin are examples of quinolones that inhibit bacterial DNA replication. These antibiotics target **DNA gyrase** and **topoisomerase IV**, enzymes responsible for unwinding and supercoiling the DNA during replication. By inhibiting these enzymes, quinolones prevent DNA replication, leading to bacterial cell death.
- **Rifamycins:** Rifampin is a rifamycin antibiotic that inhibits **RNA polymerase**, the enzyme responsible for transcribing DNA into RNA. By binding to the bacterial RNA polymerase, rifampin blocks the initiation of transcription, preventing the production of RNA and protein. Rifamycins are commonly used to treat tuberculosis and other mycobacterial infections.

12.5 INHIBITION OF FOLATE SYNTHESIS

Folate is a vital precursor for the synthesis of nucleic acids and amino acids in bacteria. Since humans obtain folate through their diet, bacteria must synthesize it themselves. Inhibiting folate synthesis in bacteria can lead to growth inhibition and cell death.

- **Sulfonamides:** Sulfonamides, such as sulfamethoxazole, inhibit the bacterial enzyme **dihydropteroate synthase**, which is involved in the early stages of folate synthesis. By preventing the synthesis of folic acid, sulfonamides disrupt DNA and protein synthesis, leading to bacteriostasis.
- **Trimethoprim:** Trimethoprim inhibits **dihydrofolate reductase**, an enzyme that converts dihydrofolate to tetrahydrofolate, a form of folate that is necessary for nucleotide synthesis. When used in combination with sulfonamides, trimethoprim forms a synergistic effect, leading to a bactericidal action.

12.6 TARGETING BACTERIAL VIRULENCE FACTORS

Beyond inhibiting growth and replication, some antibiotics and synthetic agents specifically target bacterial virulence factors, which are molecules that help bacteria cause disease. These molecules include toxins, adhesins, and enzymes involved in immune evasion.

- **Bacterial Toxin Inhibition**: Some antibiotics, such as the novel class of **immune-modulating agents**, inhibit the production or activity of bacterial toxins. For example, **clindamycin** can suppress the expression of certain toxins produced by *Staphylococcus aureus* and *Streptococcus pyogenes*, which are responsible for severe conditions like toxic shock syndrome.
- **Quorum Sensing Inhibition**: Quorum sensing is a process by which bacteria communicate with each other to coordinate their behavior, including virulence factor production. Synthetic agents that inhibit quorum sensing, such as **furanones**, are under investigation as potential anti-infective agents. These agents prevent bacterial pathogens from becoming virulent, reducing their ability to cause disease.

12.7 CHALLENGES IN THE USE OF ANTIBIOTICS AND ANTI-INFECTIVE AGENTS

While antibiotics and synthetic agents have revolutionized medicine, the emergence of antimicrobial resistance (AMR) poses significant challenges. Bacteria have evolved mechanisms to resist the effects of these drugs, such as producing enzymes that degrade antibiotics (e.g., beta-lactamases), altering drug targets, or effluxing drugs out of the bacterial cell.

The overuse and misuse of antibiotics in both healthcare and agriculture contribute significantly to the development of AMR. To combat resistance, it is essential to implement appropriate stewardship programs, improve diagnostics to ensure targeted therapy, and continue the development of novel antibiotics with new mechanisms of action.

12.8 CONCLUSION

The mechanisms of action of antibiotics and synthetic anti-infective agents are diverse and complex, allowing for targeted interventions against various microbial infections. From inhibiting cell wall synthesis to interfering with nucleic acid synthesis, these agents have saved countless lives and remain the cornerstone of modern medicine. However, the growing challenge of antimicrobial resistance calls for ongoing research, improved antibiotic stewardship, and the development of innovative therapeutic approaches. By understanding these mechanisms and applying them judiciously, the efficacy of antibiotics and synthetic anti-infective agents can be preserved for future generations.

CHAPTER 13

BACTERIAL RESISTANCE TO ANTIBIOTICS

Antibiotic resistance is one of the most significant challenges in modern medicine. Bacteria evolve mechanisms to resist the effects of antibiotics, rendering standard treatments less effective and complicating the management of infections. Understanding the factors behind bacterial resistance, how it develops, and the consequences of widespread resistance is essential to tackling this global health threat.

13.1 UNDERSTANDING ANTIBIOTIC RESISTANCE

Antibiotic resistance occurs when bacteria develop the ability to survive and proliferate in the presence of antibiotics that would normally inhibit or kill them. This resistance can be acquired naturally through mutations or transferred from other resistant bacteria through horizontal gene transfer. Resistance can be categorized as either intrinsic (natural) or acquired (developed due to exposure to antibiotics).

- **Intrinsic Resistance:** Some bacteria are naturally resistant to certain antibiotics due to inherent characteristics, such as the structure of their cell wall, the presence of efflux pumps, or the inability to take up certain drugs. For example, *Mycobacterium tuberculosis* is inherently resistant to many antibiotics due to its thick, waxy cell wall.
- **Acquired Resistance:** More commonly, bacteria acquire resistance through mutations in their genetic material or by obtaining resistance genes from other bacteria. This can occur rapidly, particularly in environments where antibiotics are frequently used.

13.2 MECHANISMS OF ANTIBIOTIC RESISTANCE

Bacteria use several mechanisms to resist the action of antibiotics. These mechanisms can be broadly classified into four main categories: modification or degradation of the antibiotic, alteration of the drug target, prevention of drug entry, and active expulsion of the antibiotic from the bacterial cell.

- **Enzymatic Degradation or Modification:** Some bacteria produce enzymes that degrade or modify antibiotics, rendering them ineffective.

A well-known example is the production of **beta-lactamases** by bacteria, which break down beta-lactam antibiotics like penicillin, making them ineffective against the bacteria producing these enzymes. Other enzymes, such as **aminoglycoside-modifying enzymes**, modify antibiotics, preventing them from binding to their targets.

- **Alteration of Target Sites:** Bacteria can alter the molecular target of an antibiotic so that the drug no longer binds effectively. For example, the modification of **penicillin-binding proteins (PBPs)** in *Staphylococcus aureus* confers resistance to beta-lactam antibiotics, as the drug can no longer bind to the altered PBPs to inhibit cell wall synthesis.
- **Prevention of Drug Entry:** Many bacteria have evolved changes in their outer membrane or cell wall that prevent the antibiotic from entering the cell. In Gram-negative bacteria, alterations in porin channels can reduce the permeability of the bacterial membrane to antibiotics like carbapenems, making them less effective.
- **Efflux Pumps:** Bacteria can also use efflux pumps to actively pump antibiotics out of the cell before they can have an effect. These pumps, which are found in the cell membrane, expel a variety of antimicrobial agents, including tetracyclines and fluoroquinolones, thus reducing the drug's intracellular concentration.

13.3 FACTORS CONTRIBUTING TO THE DEVELOPMENT OF ANTIBIOTIC RESISTANCE

The overuse and misuse of antibiotics in humans, animals, and agriculture are the primary drivers of antibiotic resistance. When antibiotics are used inappropriately or excessively, bacteria are exposed to selective pressure, which accelerates the development of resistance.

- **Overuse in Healthcare:** Antibiotic overuse in healthcare settings, such as prescribing antibiotics for viral infections (where they are ineffective), contributes to the selection of resistant bacteria. Incomplete courses of antibiotics, where patients fail to finish the full prescription, also allow surviving bacteria to adapt and develop resistance.
- **Use in Agriculture and Livestock:** The use of antibiotics in agriculture, particularly for growth promotion in livestock, contributes to the emergence of resistant bacteria. These resistant bacteria can enter the human food chain through consumption of animal products or direct contact with animals.
- **Poor Infection Control Practices:** Inadequate infection control measures in healthcare settings, such as improper hygiene, insufficient

sterilization, and improper use of medical devices, can also facilitate the spread of resistant bacteria.

- **Global Travel and Trade:** The movement of people, animals, and goods across borders contributes to the global spread of antibiotic-resistant bacteria. Resistant bacteria can easily be transmitted between countries, leading to the dissemination of resistance worldwide.

13.4 CONSEQUENCES OF ANTIBIOTIC RESISTANCE

The rise of antibiotic resistance poses a serious threat to public health, leading to longer hospital stays, increased healthcare costs, and a higher mortality rate due to infections that are no longer treatable with conventional antibiotics. Some of the most significant consequences of antibiotic resistance include:

- **Treatment Failure:** Infections caused by resistant bacteria are harder to treat, and standard antibiotics may no longer be effective. This can lead to treatment failure, prolonged illness, and, in severe cases, death. For example, infections with methicillin-resistant *Staphylococcus aureus* (MRSA) are more difficult to treat than infections with non-resistant strains.
- **Increased Healthcare Costs:** Resistant infections require more expensive and prolonged treatments, leading to increased healthcare costs. Patients may need to be treated with second- or third-line antibiotics, which are often more expensive and associated with more side effects.
- **Increased Mortality and Morbidity:** The lack of effective treatment options for resistant infections can lead to higher rates of morbidity and mortality. According to the World Health Organization (WHO), antimicrobial resistance is one of the leading threats to global health, with tens of thousands of deaths occurring annually due to resistant infections.

13.5 STRATEGIES TO COMBAT ANTIBIOTIC RESISTANCE

Efforts to combat antibiotic resistance require a multifaceted approach, involving healthcare providers, policymakers, researchers, and the general public. Key strategies to address the growing threat of antibiotic resistance include:

- **Antibiotic Stewardship:** Implementing antibiotic stewardship programs in healthcare settings is crucial for ensuring the appropriate use of antibiotics. These programs involve educating healthcare providers on when and how to prescribe antibiotics appropriately, as well as monitoring and restricting the use of antibiotics to prevent overuse and misuse.

- **Infection Control**: Strengthening infection control measures in healthcare settings can prevent the spread of resistant bacteria. Practices such as hand hygiene, isolation of infected patients, and proper sterilization of medical equipment can help reduce the transmission of resistant pathogens.
- **Public Awareness**: Educating the public about the dangers of antibiotic misuse is essential for reducing demand for unnecessary prescriptions. Public campaigns should emphasize the importance of completing prescribed antibiotic courses and the risks of self-medication and sharing antibiotics.
- **Research and Development**: Continued research into new antibiotics and alternative therapies is critical to staying ahead of resistance. The development of novel antibiotics, vaccines, and diagnostic tools can help provide new options for treating resistant infections. Additionally, research into alternatives such as phage therapy and antimicrobial peptides holds promise in the fight against resistance.
- **Regulation in Agriculture:** Reducing the use of antibiotics in agriculture, particularly for non-therapeutic purposes, is essential in controlling the spread of resistant bacteria. Some countries have already implemented regulations to restrict the use of antibiotics in livestock, but more comprehensive global regulations are needed.

13.6 CONCLUSION

Antibiotic resistance is a growing global health crisis that threatens the effectiveness of modern medicine. The development of resistance occurs through various mechanisms, and its spread is driven by overuse, misuse, and inadequate infection control. The consequences of antibiotic resistance are severe, leading to higher treatment costs, prolonged illness, and increased mortality. Combating antibiotic resistance requires a comprehensive approach that includes proper antibiotic stewardship, infection control, public awareness, research into new treatments, and regulation in agriculture. Only through coordinated efforts can we mitigate the impact of antibiotic resistance and preserve the efficacy of antibiotics for future generations.

CHAPTER 14

CLINICAL USES OF ANTIMICROBIAL DRUGS

Antimicrobial drugs are a vital part of modern medicine, playing a central role in the treatment of infections caused by microorganisms such as bacteria, viruses, fungi, and parasites. These drugs have revolutionized the way healthcare providers manage infectious diseases, significantly improving patient outcomes and reducing mortality. The clinical uses of antimicrobial drugs span a wide range of infections, each with its specific approach and targeted therapy.

14.1 OVERVIEW OF ANTIMICROBIAL DRUGS

Antimicrobial drugs encompass a broad category of medications designed to combat microbial infections. These drugs can be categorized based on the type of microorganism they target: antibiotics for bacteria, antivirals for viruses, antifungals for fungi, and antiparasitics for parasites. The development of antimicrobial drugs has made it possible to treat a variety of infections that were once life-threatening, such as tuberculosis, pneumonia, and HIV/AIDS.

Antimicrobial drugs can be further classified based on their mechanism of action, such as:

- **Bactericidal**: These drugs kill bacteria directly, such as penicillins and cephalosporins.
- **Bacteriostatic**: These drugs inhibit the growth and reproduction of bacteria without killing them, such as tetracyclines and sulfonamides.

Antimicrobial drugs can also be classified based on their spectrum of activity:

- **Narrow-Spectrum**: These drugs target a specific group of bacteria or pathogens, such as penicillin, which is effective primarily against Gram-positive bacteria.
- **Broad-Spectrum**: These drugs are effective against a wide variety of bacteria, including both Gram-positive and Gram-negative organisms, such as amoxicillin and tetracyclines.

14.2 CLINICAL USES OF ANTIBIOTICS

Antibiotics are perhaps the most well-known type of antimicrobial drugs and are primarily used to treat bacterial infections. The clinical uses of antibiotics are vast, and they are essential in managing a wide range of infections, from mild to severe.

- **Respiratory Infections:** Antibiotics are commonly used to treat respiratory tract infections, such as pneumonia, bronchitis, and sinusitis. For example, *Amoxicillin* is frequently prescribed for bacterial pneumonia, while *Azithromycin* is often used for bronchitis.
- **Urinary Tract Infections (UTIs):** UTIs are among the most common infections treated with antibiotics. Drugs like *Trimethoprim-sulfamethoxazole* and *Nitrofurantoin* are commonly used to treat uncomplicated UTIs, while more potent antibiotics, such as *Ciprofloxacin*, may be used for complicated cases.
- **Skin and Soft Tissue Infections:** Antibiotics are also widely used to treat infections of the skin and soft tissues, including cellulitis, abscesses, and wound infections. Common antibiotics for skin infections include *Cephalexin* and *Clindamycin.*
- **Gastrointestinal Infections:** Antibiotics are used to treat certain bacterial gastrointestinal infections, such as *Salmonella* and *Clostridium difficile* infections. *Metronidazole* and *Vancomycin* are examples of antibiotics used to treat C. difficile infections.
- **Sexually Transmitted Infections (STIs):** Antibiotics play a critical role in treating STIs, including *Chlamydia trachomatis* and *Neisseria gonorrhoeae* infections. *Azithromycin* and *Ceftriaxone* are commonly used to treat these infections.
- **Sepsis:** Sepsis is a life-threatening condition caused by a severe bacterial infection. Broad-spectrum antibiotics, such as *Piperacillin-tazobactam* or *Meropenem*, are often used in the treatment of sepsis to target a wide range of pathogens until the specific causative organism can be identified.

14.3 CLINICAL USES OF ANTIVIRAL DRUGS

Antiviral drugs are used to treat infections caused by viruses. Unlike antibiotics, which target bacteria, antiviral drugs specifically target viral replication, preventing the virus from multiplying and spreading. The use of antiviral drugs has greatly improved the treatment of viral infections, particularly in immunocompromised patients.

- **HIV/AIDS:** Antiretroviral therapy (ART) is the cornerstone of HIV treatment. ART involves a combination of drugs, such as *Efavirenz, Tenofovir*, and *Lamivudine*, which work together to inhibit different stages of the HIV life cycle. These drugs do not cure HIV, but they reduce the viral load and improve the patient's quality of life.

- **Influenza:** The treatment of influenza can be enhanced with antiviral drugs such as *Oseltamivir* (Tamiflu) and *Zanamivir* (Relenza). These drugs are most effective when administered early in the course of the illness and can reduce the severity and duration of symptoms.
- **Herpes Simplex Virus (HSV):** Antiviral drugs such as *Acyclovir, Valacyclovir*, and *Famciclovir* are used to treat herpes simplex virus infections, including cold sores, genital herpes, and herpes zoster (shingles). These medications help reduce the frequency of outbreaks and can suppress viral replication.
- **Hepatitis:** Chronic infections caused by hepatitis B and C viruses can be managed with antiviral medications. *Tenofovir* and *Entecavir* are used for hepatitis B, while *Sofosbuvir* and *Daclatasvir* are used for hepatitis C to prevent the progression of liver disease and reduce the risk of liver cancer.
- **COVID-19:** The emergence of the COVID-19 pandemic has highlighted the importance of antiviral drugs in managing viral infections. Drugs like *Remdesivir* and *Molnupiravir* have been used to treat COVID-19 by inhibiting viral replication, although their efficacy continues to be studied.

14.4 CLINICAL USES OF ANTIFUNGAL DRUGS

Fungal infections are another class of infections that can be treated with antimicrobial drugs. While fungal infections are less common than bacterial infections, they can be serious, especially in immunocompromised individuals. Antifungal drugs are classified based on their mechanism of action and spectrum of activity.

- **Superficial Fungal Infections:** Superficial fungal infections, such as athlete's foot, ringworm, and candidiasis (thrush), are often treated with topical antifungal agents. Commonly used antifungals include *Clotrimazole* and *Ketoconazole* for skin infections, and *Nystatin* for oral candidiasis.
- **Systemic Fungal Infections:** More serious systemic fungal infections, such as histoplasmosis, aspergillosis, and cryptococcosis, require systemic treatment with drugs like *Amphotericin B, Fluconazole*, and *Itraconazole*. These drugs are effective at treating infections that affect internal organs or are widespread in the body.

- **Candidemia and Fungal Meningitis:** For life-threatening systemic infections like candidemia and fungal meningitis, intravenous antifungal drugs such as *Voriconazole* and *Echinocandins* (e.g., *Caspofungin*) are used to target the infection and prevent complications.

14.5 CLINICAL USES OF ANTIPARASITIC DRUGS

Parasitic infections, caused by protozoa, helminths, and ectoparasites, can also be treated with antimicrobial agents. Antiparasitic drugs are used in the treatment of diseases such as malaria, leishmaniasis, and helminth infections.

- **Malaria:** Malaria, caused by the *Plasmodium* species, is one of the most well-known parasitic diseases. Antimalarial drugs such as *Chloroquine, Artemisinin-based combination therapies (ACTs),* and *Mefloquine* are commonly used to treat and prevent malaria.
- **Helminth Infections:** Helminth infections, including roundworm, hookworm, and tapeworm infections, are treated with drugs such as *Mebendazole, Albendazole,* and *Praziquantel.* These drugs kill the parasites or inhibit their growth, providing relief from symptoms.
- **Leishmaniasis:** Leishmaniasis, caused by *Leishmania* species, is treated with drugs like *Miltefosine* and *Amphotericin B.* These medications help control the infection and prevent the development of more serious forms of the disease, such as visceral leishmaniasis.

14.6 CHALLENGES AND CONSIDERATIONS IN THE USE OF ANTIMICROBIAL DRUGS

While antimicrobial drugs are indispensable in modern medicine, their use is not without challenges. These challenges include:

- **Antibiotic Resistance:** The overuse and misuse of antibiotics have led to the development of antibiotic-resistant bacteria, making treatment more difficult and less effective. The growing threat of resistance has led to the search for new antibiotics and alternative therapies.
- **Side Effects:** Many antimicrobial drugs have side effects, ranging from mild symptoms like nausea to more serious complications such as liver damage or nephrotoxicity. Healthcare providers must carefully weigh the benefits and risks of treatment.
- **Drug Interactions:** Antimicrobial drugs may interact with other medications, reducing their efficacy or increasing the risk of adverse reactions. It is essential for clinicians to consider potential drug-drug interactions when prescribing these drugs.

14.7 CONCLUSION

Antimicrobial drugs are essential in the clinical management of infectious diseases, offering effective treatment options for bacterial, viral, fungal, and parasitic infections.

The proper use of these drugs is critical to ensuring patient outcomes and minimizing the impact of antimicrobial resistance. As medical research continues to advance, new and improved antimicrobial agents are being developed, offering hope for the future of infection management.

However, their use must be carefully monitored to prevent misuse and the emergence of resistant organisms, ensuring that these life-saving drugs remain effective for generations to come.

CHAPTER 15

ANTIBIOTIC PRESCRIBING AND ANTIMICROBIAL STEWARDSHIP

Antibiotics have revolutionized the treatment of bacterial infections, drastically reducing mortality rates and improving patient outcomes. However, inappropriate or excessive use of antibiotics has led to a growing problem of antibiotic resistance, which poses a serious public health threat. In response, antibiotic prescribing and antimicrobial stewardship programs (ASPs) have become essential strategies in promoting the responsible use of antibiotics, optimizing treatment outcomes, and combating resistance.

15.1 UNDERSTANDING ANTIBIOTIC PRESCRIBING

Antibiotic prescribing refers to the practice of healthcare providers recommending antibiotics to treat infections caused by bacteria. Proper antibiotic prescribing involves selecting the most appropriate antibiotic, at the correct dose, and for the appropriate duration to effectively treat the infection while minimizing the risk of side effects and resistance development.

Antibiotic prescribing should be based on a variety of factors, including:

- **Diagnosis:** Accurate diagnosis is crucial for determining whether an infection is bacterial in nature and whether antibiotics are needed. Many infections are caused by viruses (e.g., the common cold or influenza), against which antibiotics are ineffective. Incorrectly prescribing antibiotics for viral infections contributes to unnecessary antibiotic use and resistance.
- **Antibiotic Spectrum:** Healthcare providers should choose antibiotics that target the specific bacteria responsible for the infection. Narrow-spectrum antibiotics should be prioritized whenever possible to minimize disruption to the body's normal flora and reduce the development of resistance. Broad-spectrum antibiotics, though effective against a wide range of bacteria, should be reserved for situations where the causative organism is unknown or when infections involve multiple pathogens.
- **Patient Factors:** Consideration of the patient's medical history, allergies, kidney and liver function, and other underlying conditions is

essential when prescribing antibiotics. Some patients may have specific needs or contraindications for certain antibiotics, and tailoring the treatment plan can help reduce adverse reactions.

- **Antibiotic Resistance Patterns:** Knowledge of local resistance patterns can inform the choice of antibiotic. In some regions, certain bacteria may be resistant to commonly prescribed antibiotics, and alternative therapies may be necessary. This is especially important in healthcare settings, where multidrug-resistant organisms are more common.

15.2 THE IMPORTANCE OF ANTIMICROBIAL STEWARDSHIP

Antimicrobial stewardship (AMS) refers to coordinated efforts to promote the appropriate use of antibiotics and other antimicrobial agents. The main goal of AMS is to optimize the treatment of infections while minimizing the risks of resistance, adverse effects, and unnecessary treatments. Effective stewardship involves a collaborative approach across healthcare systems and aims to improve clinical outcomes and patient safety while ensuring the longevity of existing antibiotics.

Key principles of antimicrobial stewardship include:

- **Appropriate Antibiotic Use:** Ensuring antibiotics are used only when necessary and that the correct antibiotic is selected based on the infection type and susceptibility of the microorganism. This involves avoiding unnecessary prescriptions, reducing the duration of therapy when possible, and choosing narrow-spectrum agents over broad-spectrum antibiotics.
- **Monitoring and Feedback:** Monitoring antibiotic use and providing feedback to clinicians on prescribing patterns is essential for improving practices. This can involve auditing prescribing data and discussing findings with healthcare providers to encourage better decision-making.
- **Education and Training:** Education is a cornerstone of effective antimicrobial stewardship. Healthcare professionals need to be continually educated about the risks of antibiotic overuse, resistance mechanisms, and the best practices for antibiotic prescribing. Regular training sessions and updates on local resistance patterns can support clinicians in making informed choices.
- **Infection Prevention and Control:** Prevention strategies such as vaccination, hygiene practices, and infection control measures are essential components of antimicrobial stewardship. Preventing infections in the first place reduces the need for antibiotics and limits the opportunities for resistance to develop.

- **Treatment Optimization:** AMS encourages regular review of ongoing antibiotic therapies, particularly in hospitalized patients. Clinicians are encouraged to reassess the need for continued antibiotic therapy, de-escalating to narrower-spectrum agents or stopping antibiotics when no longer necessary.

15.3 ANTIBIOTIC RESISTANCE AND ITS CONSEQUENCES

Antibiotic resistance occurs when bacteria evolve mechanisms that allow them to survive exposure to drugs that would normally kill them or inhibit their growth. This can happen through mutations in bacterial DNA or by acquiring resistance genes from other bacteria. Over time, resistant bacteria can spread, rendering certain antibiotics ineffective and making infections harder to treat.

The consequences of antibiotic resistance are far-reaching:

- **Increased Morbidity and Mortality:** Infections caused by resistant bacteria are often more difficult to treat, leading to prolonged illnesses, complications, and higher mortality rates. For example, infections caused by multidrug-resistant organisms, such as *Methicillin-resistant Staphylococcus aureus* (MRSA), can lead to sepsis and organ failure if not treated promptly with effective antibiotics.
- **Longer Hospital Stays and Higher Healthcare Costs:** Resistant infections often require longer treatments, more intensive care, and the use of more expensive drugs. This leads to longer hospital stays and an increase in overall healthcare costs, placing a financial strain on both healthcare systems and patients.
- **Limited Treatment Options:** As resistance continues to develop, fewer antibiotics remain effective against certain pathogens. This limits treatment options and leaves healthcare providers with few alternatives for managing infections, particularly in vulnerable populations, such as immunocompromised individuals and the elderly.
- **Global Health Threat:** Antibiotic resistance is a global issue, as resistant bacteria can spread across borders through travel and trade. Without coordinated international efforts to curb resistance, infections that were once treatable with antibiotics could become untreatable, leading to an increase in preventable deaths worldwide.

15.4 STRATEGIES FOR EFFECTIVE ANTIBIOTIC PRESCRIBING

To combat the growing issue of antibiotic resistance, several strategies have been proposed to optimize antibiotic prescribing:

- **Adherence to Guidelines:** National and international guidelines provide evidence-based recommendations on the appropriate use of antibiotics for various infections. Healthcare providers should be familiar with these guidelines and use them as a framework for prescribing antibiotics.
- **Rapid Diagnostic Tests:** The use of rapid diagnostic tests (RDTs) can help identify the causative pathogen more quickly, allowing for more accurate antibiotic prescribing. For example, molecular tests can detect bacterial DNA and identify specific pathogens or resistance genes in real-time, reducing the time spent on empirical treatment and decreasing the likelihood of unnecessary antibiotic use.
- **Antibiotic Stewardship Committees:** Establishing multidisciplinary antibiotic stewardship committees within healthcare organizations is crucial. These committees include infectious disease specialists, microbiologists, pharmacists, and other healthcare professionals who collaborate to ensure the appropriate use of antibiotics across the organization.
- **Delaying Antibiotic Use:** In cases where a bacterial infection is suspected but not confirmed, a "watchful waiting" approach may be appropriate. Delaying antibiotic treatment for a short period allows the body's immune system time to fight the infection, reducing unnecessary antibiotic use.
- **Antibiotic Stewardship in Outpatient Settings:** Antibiotic stewardship is not limited to hospitals and healthcare facilities. In outpatient settings, healthcare providers must be vigilant in their prescribing practices, particularly for common conditions like sore throats, sinusitis, and urinary tract infections, where antibiotics are often overprescribed.

15.5 CHALLENGES TO IMPLEMENTING ANTIMICROBIAL STEWARDSHIP

While antimicrobial stewardship programs have proven effective in many settings, several challenges remain in their implementation:

- **Resistance to Change:** Healthcare providers may be resistant to changing their prescribing habits, especially if they are accustomed to treating certain conditions with antibiotics. Overcoming these attitudes requires ongoing education and evidence-based data that demonstrate the benefits of appropriate antibiotic use.
- **Time Constraints:** Clinicians often face time constraints in busy healthcare settings, making it difficult to thoroughly review antibiotic

use or consult stewardship guidelines. Streamlining workflows and providing decision support tools can help make antibiotic stewardship a seamless part of daily practice.

- **Inadequate Diagnostic Tools:** In many cases, rapid diagnostic tests are not available or not widely implemented, leading to the continued reliance on broad-spectrum antibiotics as a first-line treatment. The development and adoption of faster, more affordable diagnostic tools could help address this limitation.
- **Resource Limitations:** Implementing robust antimicrobial stewardship programs requires significant resources, including trained personnel, information technology infrastructure, and funding. In low-resource settings, these barriers may prevent the widespread adoption of AMS initiatives.

15.6 CONCLUSION

Antibiotic prescribing and antimicrobial stewardship are critical components of modern healthcare. While antibiotics remain one of the most important tools for treating bacterial infections, their overuse and misuse have led to the growing challenge of antibiotic resistance. Antimicrobial stewardship programs play a central role in optimizing antibiotic use, improving patient outcomes, and preserving the effectiveness of antibiotics for future generations. Through collaborative efforts, continued education, and adherence to best practices, healthcare providers can help mitigate the impact of antibiotic resistance and safeguard the future of antimicrobial therapies.

CHAPTER 16

INFECTION PREVENTION AND CONTROL: HEALTHCARE-ASSOCIATED INFECTION

Healthcare-associated infections (HAIs) are infections that patients acquire while receiving medical treatment within a healthcare setting, such as hospitals, clinics, or nursing homes. These infections pose significant risks to patients, healthcare workers, and the overall healthcare system. They result in extended hospital stays, increased healthcare costs, and in some cases, severe complications or death. The importance of infection prevention and control (IPC) within healthcare settings is paramount to safeguarding patient health, reducing the burden on healthcare systems, and improving clinical outcomes.

16.1 UNDERSTANDING HEALTHCARE-ASSOCIATED INFECTIONS (HAIS)

Healthcare-associated infections are infections that patients acquire as a result of their interactions with the healthcare system, either while undergoing medical treatments, surgeries, or as a consequence of hospitalization. These infections can be caused by a wide variety of pathogens, including bacteria, viruses, fungi, and parasites. Common types of HAIs include urinary tract infections (UTIs), surgical site infections (SSIs), bloodstream infections (BSIs), pneumonia, and gastrointestinal infections.

HAIs can occur through various routes, such as:

- **Direct Contact**: Healthcare workers, patients, or visitors can transmit pathogens through physical contact. This includes touching contaminated surfaces, medical equipment, or handling open wounds.
- **Airborne Transmission**: Infections like tuberculosis or respiratory viruses can spread through the air, especially in crowded or poorly ventilated spaces.
- **Contaminated Medical Equipment**: Non-sterile medical instruments, devices, and equipment, such as catheters, ventilators, and intravenous lines, can transmit infections.

- **Improper Handling of Bodily Fluids**: Blood, urine, feces, or respiratory secretions that are not properly handled or disposed of can become vectors for infection.

Infections may develop within a few hours to days after exposure and can often lead to complications such as sepsis, multi-organ failure, and even death.

16.2 IMPACT OF HEALTHCARE-ASSOCIATED INFECTIONS

The impact of HAIs on patients, healthcare providers, and the healthcare system is profound.

- **Patient Outcomes**: HAIs are associated with prolonged hospitalization, increased medical complications, and higher rates of mortality. Vulnerable populations, such as the elderly, immunocompromised individuals, and those undergoing major surgeries, are particularly at risk. The quality of life of patients can also be significantly diminished by infections acquired in the hospital, and recovery times are often extended.
- **Healthcare Costs**: HAIs significantly increase healthcare costs due to longer hospital stays, more intensive treatments, and the need for additional medications and interventions. Studies have shown that HAIs contribute to billions of dollars in extra healthcare expenses each year. Hospitals are required to bear the financial burden of treating patients who acquire infections within their facilities.
- **Antimicrobial Resistance (AMR)**: One of the most alarming consequences of HAIs is the rise of antimicrobial resistance. Inadequate infection control measures can lead to the spread of drug-resistant bacteria, making infections harder to treat and increasing the risk of complications. The misuse and overuse of antibiotics in hospital settings are primary contributors to AMR, complicating the management of both HAIs and community-acquired infections.
- **Healthcare Worker Safety**: Healthcare workers are at risk of contracting infections from patients, especially when proper protective equipment and safety protocols are not followed. This not only impacts the health of healthcare workers but also increases the likelihood of cross-contamination between patients and staff.

16.3 INFECTION PREVENTION AND CONTROL (IPC) STRATEGIES

Effective infection prevention and control strategies are essential in minimizing the incidence of HAIs. A comprehensive IPC program should include a combination of preventive practices, surveillance systems, and policies that promote cleanliness, sterilization, and proper hygiene practices throughout healthcare facilities.

16.3.1 Hand Hygiene

Hand hygiene is the most fundamental and effective way to prevent the transmission of infections in healthcare settings. Healthcare workers, including doctors, nurses, and other staff, must regularly wash their hands with soap and water or use alcohol-based hand sanitizers. Hand hygiene should be performed before and after patient contact, before performing medical procedures, after handling contaminated materials, and after using the restroom. Adherence to hand hygiene protocols is critical in reducing the spread of harmful microorganisms, such as Clostridium difficile, MRSA, and norovirus, which can be transmitted from patient to patient via contaminated hands.

16.3.2 Personal Protective Equipment (PPE)

Personal protective equipment, including gloves, gowns, masks, and face shields, helps create a barrier between healthcare workers and infectious agents. PPE should be worn when there is a potential for exposure to blood, bodily fluids, or contaminated surfaces. The proper use of PPE prevents the spread of infections from patient to healthcare worker and vice versa. Moreover, PPE helps protect patients from pathogens carried by healthcare workers. Training healthcare workers on the proper use, disposal, and maintenance of PPE is crucial to ensure its effectiveness.

16.3.3 Environmental Cleaning and Disinfection

Proper environmental cleaning and disinfection are essential in minimizing the risk of HAIs. Hospital surfaces, including patient beds, medical equipment, doorknobs, and high-touch areas, can harbor pathogens for extended periods. Routine cleaning with disinfectants and regular sterilization of medical instruments is necessary to ensure a hygienic environment. Healthcare facilities should have protocols in place for cleaning and disinfecting patient rooms, bathrooms, and common areas. Special attention should be given to high-risk areas such as intensive care units (ICUs) and operating rooms, where the potential for pathogen transmission is higher.

16.3.4 Sterilization and Disinfection of Medical Equipment

Medical equipment, such as catheters, ventilators, and endoscopes, must be properly sterilized or disinfected after each use. Non-sterile instruments can introduce infections directly into the body, leading to severe complications, including bloodstream infections and pneumonia. Sterilization protocols must follow strict guidelines, including the use of autoclaves, heat, or chemical disinfectants, depending on the type of equipment and the level of sterilization required. Healthcare facilities must ensure that medical equipment is reprocessed in compliance with infection control standards.

16.3.5 Surveillance and Early Detection

Surveillance plays a crucial role in identifying the emergence of HAIs and monitoring trends over time. Healthcare facilities should have a surveillance system in place to track infection rates, detect outbreaks, and identify potential risk factors associated with HAIs. Early detection allows healthcare providers to intervene quickly, isolate infected patients, and prevent the further spread of infections. For example, rapid diagnostic tests can help identify bacterial infections in a timely manner, enabling healthcare workers to implement appropriate isolation protocols and targeted antimicrobial treatments.

16.3.6 Antimicrobial Stewardship

The responsible use of antibiotics is a key element of infection prevention. Overuse or misuse of antibiotics leads to the development of antimicrobial-resistant organisms, making infections harder to treat and increasing the risk of HAIs. Antimicrobial stewardship programs aim to optimize the use of antibiotics by promoting the correct choice, dosage, and duration of therapy. These programs also help reduce the over-prescribing of antibiotics for viral infections, such as the common cold or influenza, where antibiotics are ineffective.

16.3.7 Isolation Precautions

For patients diagnosed with contagious infections, isolation precautions must be implemented to prevent the transmission of pathogens to other patients and healthcare workers. Isolation measures may include placing patients in private rooms, using dedicated medical equipment, and following strict PPE guidelines when interacting with infected patients. Isolation protocols vary depending on the infectious agent. For example, contact precautions are necessary for patients with drug-resistant infections, while droplet or airborne precautions may be required for patients with respiratory infections such as tuberculosis.

16.4 CHALLENGES IN INFECTION PREVENTION AND CONTROL

Despite the importance of infection prevention and control measures, several challenges persist in healthcare settings:

- **Compliance and Training:** Ensuring consistent adherence to IPC protocols by healthcare workers is challenging. Busy work environments, lack of training, and forgetfulness can lead to lapses in hygiene practices and failure to follow proper protocols.
- **Resource Limitations:** Some healthcare facilities, particularly those in low-income regions, may lack the resources to implement and maintain robust IPC programs. Limited access to cleaning supplies, PPE, and diagnostic tools can hinder infection control efforts.

- **Antimicrobial Resistance:** The rise of antimicrobial resistance adds a layer of complexity to infection prevention. As resistant pathogens become more common, healthcare providers must employ additional strategies to manage infections and prevent outbreaks.
- **Patient Education:** Patients themselves play a role in infection prevention. Educating patients about the importance of hand hygiene, adherence to treatment plans, and the proper use of medical devices can help reduce the risk of HAIs.

16.5 CONCLUSION

Healthcare-associated infections represent a significant challenge to patient safety and healthcare delivery worldwide. However, with robust infection prevention and control measures in place, the risk of HAIs can be greatly reduced. A comprehensive approach that includes hand hygiene, the use of PPE, environmental cleaning, proper sterilization of medical equipment, surveillance, antimicrobial stewardship, and isolation precautions is essential to minimizing the impact of infections. Effective infection prevention requires the collaboration of healthcare workers, administrators, patients, and public health authorities to ensure that healthcare environments remain safe and conducive to healing.

PART 5:

CONTAMINATION AND CONTAMINATION CONTROL

CHAPTER 17

MICROBIAL SPOILAGE, INFECTION RISK AND CONTAMINATION CONTROL

Microbial spoilage, infection risk, and contamination control are interconnected concepts that play a pivotal role in various industries, particularly in food, pharmaceuticals, healthcare, and biotechnology. These processes are critical in maintaining product quality, ensuring public health, and minimizing the spread of infectious agents. Microbial spoilage occurs when microorganisms, such as bacteria, molds, and yeasts, degrade products, causing loss of quality, safety, and nutritional value. Infection risks, meanwhile, arise when harmful microorganisms invade the body, leading to illness. Contamination control is the practice of minimizing or eliminating the presence of harmful microorganisms in the production, storage, and distribution of goods.

17.1 MICROBIAL SPOILAGE AND ITS CAUSES

Microbial spoilage refers to the deterioration of food, pharmaceuticals, or other products caused by microorganisms, which thrive in favorable conditions, such as moisture, warmth, and nutrients. In the context of food, spoilage can result in off-flavors, discoloration, undesirable textures, and the potential for pathogenic organisms to grow.

Several factors contribute to microbial spoilage, including:

- **Types of Microorganisms:** Bacteria, molds, and yeasts are the primary culprits behind microbial spoilage. Bacteria such as *Salmonella*, *E. coli*, and *Listeria* are responsible for causing foodborne illnesses. Meanwhile, molds and yeasts contribute to the physical degradation of products, especially in foods like bread, cheese, and fruit.
- **Environmental Factors:** Temperature, humidity, and oxygen availability are key environmental conditions that influence microbial growth. For example, higher temperatures often accelerate bacterial growth, while moisture promotes the growth of molds and yeasts. Poor storage conditions, improper handling, and contamination during processing all contribute to microbial spoilage.

- **Substrate Types:** Different food or pharmaceutical products provide ideal environments for specific types of microorganisms. For instance, protein-rich foods like meat and dairy products are more susceptible to bacterial spoilage, while sugar-rich foods like fruit and syrups attract yeast growth. Similarly, certain pharmaceutical formulations, like liquid-based medicines, are prone to microbial contamination if not stored properly.

Microbial spoilage is often accompanied by a decline in the product's sensory attributes, including taste, smell, texture, and appearance. This not only compromises consumer satisfaction but also poses a health risk when the microorganisms responsible for spoilage are pathogenic.

17.2 INFECTION RISKS IN HEALTHCARE AND FOOD INDUSTRIES

The risk of infection is a direct consequence of microbial contamination, and it is a significant concern in both healthcare settings and the food industry. In healthcare, patients are at risk of infections due to exposure to harmful microorganisms present in contaminated medical equipment, surgical instruments, or the hospital environment. Healthcare-associated infections (HAIs) can occur when bacteria, viruses, or fungi enter the body through wounds, intravenous lines, or respiratory pathways. The risk of infection in the food industry also stems from microbial contamination. When food is contaminated with harmful bacteria, viruses, or parasites, it can lead to foodborne illnesses, which can be serious or even fatal. Common pathogens that cause foodborne infections include *Salmonella*, *Campylobacter*, *Norovirus*, and *Clostridium botulinum*. These microorganisms thrive in improperly stored or handled food, leading to contamination during production, processing, transportation, or preparation.

In both healthcare and food industries, the risk of infection increases when contamination goes undetected or is poorly controlled. Microbial pathogens are often invisible to the naked eye, making it difficult to identify contamination early. Additionally, infected individuals, if not isolated or treated promptly, can spread microorganisms to others, exacerbating the risk.

17.3 CONTAMINATION CONTROL IN THE FOOD INDUSTRY

Contamination control is a crucial aspect of the food industry, where microbial spoilage and infection risks are a constant concern. Effective contamination control measures prevent the growth of harmful microorganisms and ensure the safety and quality of food products.

17.3.1 Sanitation and Hygiene Practices

Sanitation is the first line of defense against microbial contamination in food processing. Proper cleaning and disinfection of food contact surfaces, utensils, equipment, and storage facilities are essential to prevent microbial growth. In addition, food handlers must follow stringent hygiene practices, including frequent handwashing, wearing gloves, and using face masks when necessary. These steps minimize the chances of cross-contamination between raw ingredients and ready-to-eat foods. Food processing plants must implement comprehensive sanitation schedules that include daily cleaning, deep cleaning, and regular audits to ensure that all areas are adequately sanitized. The use of sanitizers that are effective against a wide range of microorganisms, including bacteria, viruses, and fungi, is also crucial.

17.3.2 Temperature Control

Temperature control is another essential aspect of contamination control. Microorganisms proliferate rapidly at temperatures between 4°C and 60°C, often referred to as the "danger zone" for food safety. Maintaining food at proper temperatures during storage, transportation, and preparation helps slow down microbial growth. Refrigeration, freezing, and cooking at high temperatures are all methods of controlling microbial contamination. For example, raw meat should be refrigerated at or below 4°C, and cooked foods should be kept above 60°C to prevent bacterial growth. Regular monitoring of temperature settings and ensuring proper storage practices are fundamental to maintaining food safety.

17.3.3 Use of Preservatives and Additives

Preservatives and additives are often used to control microbial growth in food products. These substances help extend shelf life by inhibiting or slowing down the activity of spoilage microorganisms. Common preservatives include salt, sugar, vinegar, and chemicals like sodium benzoate, which help prevent the growth of bacteria, molds, and yeasts. However, the use of preservatives must be regulated to ensure their safety and effectiveness. Overuse or improper application of preservatives can lead to health risks for consumers, making it important to follow established guidelines.

17.4 CONTAMINATION CONTROL IN HEALTHCARE SETTINGS

In healthcare settings, contamination control is essential for preventing infections, particularly healthcare-associated infections (HAIs). Effective strategies ensure that patients are not exposed to pathogens, reducing the risk of illness and enhancing overall patient care.

17.4.1 Sterilization of Medical Equipment

The sterilization of medical equipment is one of the most critical aspects of contamination control in healthcare. Surgical instruments, catheters, and other medical devices must be sterilized to eliminate any microbial contamination. Autoclaving, using high-pressure steam, is the most common method for sterilizing medical equipment. Other techniques, such as ultraviolet (UV) light, chemical disinfectants, or dry heat, can also be used depending on the material and type of equipment. Proper storage of sterilized items is equally important. Medical instruments should be stored in a clean, dry environment to avoid re-contamination before use.

17.4.2 Personal Protective Equipment (PPE)

Personal protective equipment (PPE) plays a crucial role in preventing the spread of infections in healthcare settings. Gloves, gowns, masks, and face shields are used to protect healthcare workers from exposure to potentially harmful microorganisms. PPE also acts as a barrier to prevent healthcare workers from transmitting pathogens to patients. Healthcare workers must adhere to strict guidelines on when and how to use PPE. For example, gloves should be worn when handling contaminated items or performing procedures that may expose workers to bodily fluids. Masks and gowns should be used in high-risk situations, such as surgeries or when dealing with patients in isolation.

17.4.3 Infection Control Policies and Training

Infection control policies and regular training programs for healthcare workers are essential for preventing contamination. These policies should cover protocols for patient isolation, handling of infectious materials, hygiene practices, and the proper use of PPE. Healthcare workers must be continuously educated about the importance of hand hygiene, disinfection, and sterilization to minimize the risk of cross-contamination. In addition, monitoring compliance with infection control policies through audits and feedback is necessary to ensure that infection prevention efforts are effective.

17.5 CONCLUSION

Microbial spoilage, infection risk, and contamination control are fundamental issues that impact the food, healthcare, and pharmaceutical industries. Effective strategies and practices must be implemented to prevent microbial growth, protect human health, and maintain product quality. Whether through hygiene practices, temperature control, sterilization of equipment, or the use of preservatives, controlling contamination is critical to minimizing the risks posed by harmful microorganisms. Both industries must remain vigilant in their efforts to reduce contamination and safeguard public health, ensuring a safer and healthier environment for all.

CHAPTER 18

CHEMICAL DISINFECTANTS, ANTISEPTICS AND PRESERVATIVES

Chemical disinfectants, antiseptics, and preservatives play crucial roles in various sectors, from healthcare and pharmaceuticals to food and beverage industries.

These agents work to eliminate, inhibit, or reduce microbial growth, preventing infection and prolonging product shelf life. While they share common goals, each category serves unique functions suited to specific contexts.

18.1 UNDERSTANDING CHEMICAL DISINFECTANTS

Chemical disinfectants are substances applied to non-living objects and surfaces to destroy microorganisms or render them inactive. Disinfectants are vital in healthcare settings, laboratories, and food processing environments to prevent the spread of infectious diseases.

18.1.1 Types of Chemical Disinfectants

Several types of chemical disinfectants are commonly used:

- **Alcohols (e.g. Ethanol and Isopropanol)**: Alcohol-based disinfectants are effective against bacteria and many viruses. They work by denaturing proteins and dissolving lipid membranes, making them widely used in hand sanitizers and surface disinfectants. However, alcohols are less effective against spores and may require higher concentrations for optimal efficacy.
- **Chlorine Compounds (e.g. Sodium Hypochlorite)**: Chlorine is a powerful disinfectant used to treat drinking water, sanitize surfaces, and disinfect swimming pools. It works by oxidizing cellular components, effectively killing bacteria, viruses, and fungi. However, chlorine can be corrosive and may cause skin irritation upon prolonged exposure.
- **Quaternary Ammonium Compounds (Quats)**: Quats are commonly used for disinfecting surfaces in hospitals and other facilities. They work by disrupting microbial cell membranes and are particularly

effective against gram-positive bacteria. Quats are often combined with other disinfectants for enhanced efficacy.

- **Aldehydes (e.g. Glutaraldehyde)**: Aldehydes are effective against a broad range of microorganisms, including bacterial spores. They are often used in the sterilization of medical equipment. However, due to their toxicity and potential health risks, they are used only when necessary and in well-ventilated areas.

18.1.2 Applications of Disinfectants

Disinfectants play a central role in infection control by decontaminating surfaces, equipment, and facilities where microbial contamination poses a risk. In hospitals, they are essential for disinfecting surgical instruments, cleaning patient rooms, and maintaining sterile environments. In food processing, disinfectants are used to sanitize equipment, prevent contamination, and ensure the safety of food products.

18.2 THE ROLE OF ANTISEPTICS IN INFECTION PREVENTION

Antiseptics are chemical agents applied to living tissues, such as skin or mucous membranes, to prevent infection by inhibiting or killing microorganisms. Unlike disinfectants, which are used on inanimate surfaces, antiseptics are designed to be safe for use on or within the body. They are crucial for preventing wound infections, reducing microbial load on the skin before surgery, and managing minor cuts or abrasions.

18.2.1 Types of Antiseptics

Different types of antiseptics are used based on the specific need and area of application:

- **Alcohols (Ethanol and Isopropanol)**: In addition to their use as disinfectants, alcohols are widely used as antiseptics for skin disinfection before injections or surgical procedures. They work quickly to reduce bacterial populations on the skin.
- **Iodine and Iodophors (e.g. Povidone-Iodine)**: Iodine-based antiseptics are effective against a wide range of microorganisms, including bacteria, viruses, and fungi. Povidone-iodine is commonly used to prepare the skin before surgery and to disinfect wounds. It works by disrupting cellular enzymes and proteins.
- **Chlorhexidine**: Chlorhexidine is a broad-spectrum antiseptic that disrupts cell membranes and is commonly used in mouthwashes, skin disinfectants, and wound care products. Its residual activity makes it highly effective in reducing bacterial populations over extended periods.

- **Hydrogen Peroxide**: Hydrogen peroxide is an antiseptic effective against a range of pathogens. It works by releasing oxygen upon contact with tissues, creating an environment that is toxic to anaerobic bacteria. It is frequently used to clean minor cuts and abrasions.

18.2.2 Applications of Antiseptics

Antiseptics are vital in healthcare for reducing the risk of infections associated with medical procedures. They are used for preoperative skin preparation, hand sanitization, wound care, and in products like mouthwashes to reduce oral bacteria. In home and community settings, antiseptics are found in first aid kits, enabling individuals to treat minor injuries promptly and safely.

18.3 PRESERVATIVES: EXTENDING SHELF LIFE AND PREVENTING SPOILAGE

Preservatives are substances added to food, pharmaceuticals, cosmetics, and other perishable products to prevent microbial spoilage and prolong shelf life. While preservatives may not entirely kill microorganisms, they inhibit their growth, ensuring product quality and safety over time.

18.3.1 Types of Preservatives

Preservatives can be classified into natural and synthetic categories, with each type suited to specific applications:

- **Natural Preservatives (e.g. Salt, Vinegar, and Sugar)**: Natural preservatives have been used for centuries to prevent spoilage. Salt and sugar work by drawing moisture out of cells, creating an environment where bacteria cannot thrive. Vinegar, an acidic substance, inhibits bacterial growth by lowering pH levels.
- **Chemical Preservatives (e.g. Benzoates, Sorbates, and Sulfites)**: These preservatives are synthetic compounds that inhibit microbial growth in various products. Benzoates and sorbates are commonly used in foods, while sulfites are often added to beverages. Chemical preservatives work by disrupting microbial metabolic processes, effectively preventing spoilage.
- **Antioxidants (e.g. BHT and BHA)**: While primarily used to prevent oxidation, antioxidants also inhibit the growth of certain microorganisms, indirectly preserving food quality. They are commonly found in processed foods, oils, and cosmetic products.

18.3.2 Applications of Preservatives

Preservatives are essential in food safety and quality management. In food processing, they are used to extend the shelf life of products such as baked goods, dairy, and canned foods. In pharmaceuticals, preservatives prevent microbial contamination in formulations like syrups, eye drops, and creams. Cosmetics and personal care products also use preservatives to maintain their stability and prevent skin infections caused by contaminated products.

18.4 MECHANISMS OF ACTION OF DISINFECTANTS, ANTISEPTICS, AND PRESERVATIVES

The mechanisms of action for disinfectants, antiseptics, and preservatives vary based on their chemical composition, application, and target microorganisms. Generally, these agents work by damaging cellular structures, disrupting metabolic pathways, or creating an environment unsuitable for microbial survival.

- **Cell Membrane Disruption:** Many disinfectants and antiseptics, such as alcohols, work by disrupting the lipid membranes of microbial cells. This causes cell contents to leak out, leading to cell death. Quats and chlorhexidine also target cell membranes, making them effective against a broad range of bacteria.
- **Protein Denaturation:** Agents like alcohols, iodine, and aldehydes denature proteins within microbial cells. By disrupting enzymes and structural proteins, these agents halt essential cellular processes, effectively killing the microorganisms.
- **Oxidative Damage:** Oxidizing agents, such as hydrogen peroxide and chlorine compounds, release reactive oxygen species that damage microbial DNA, proteins, and lipids. This oxidative stress inhibits microbial function and leads to cell death.
- **pH Alteration:** Certain preservatives, like acetic acid (vinegar) and benzoic acid, work by lowering the pH of the product. Many microorganisms cannot grow or survive in acidic conditions, making pH alteration an effective preservation method.

18.5 CHALLENGES IN USING CHEMICAL AGENTS

While chemical disinfectants, antiseptics, and preservatives are essential for controlling microbial growth, they come with certain challenges:

- **Resistance Development:** Over time, microorganisms can develop resistance to chemical agents, rendering them less effective. For example, bacteria exposed to sub-lethal levels of disinfectants may adapt and become less susceptible, leading to challenges in healthcare and food safety.
- **Toxicity and Safety Concerns:** Some chemical agents, like aldehydes,

can be toxic to humans, causing respiratory or skin irritation. Proper handling and use guidelines are essential to prevent harmful side effects and ensure safety.

- **Environmental Impact:** Many chemical agents, particularly chlorine-based disinfectants, can harm aquatic ecosystems if not disposed of properly. The environmental implications of these agents necessitate the development of eco-friendly alternatives and waste management practices.

18.6 FUTURE DIRECTIONS AND INNOVATIONS

The demand for safe, effective, and sustainable disinfection and preservation methods has led to ongoing research into new chemical agents and technologies:

- **Biodegradable Preservatives:** Researchers are developing biodegradable preservatives derived from natural sources, which reduce environmental impact while maintaining effectiveness. Plant-derived essential oils and extracts, for instance, have shown promise as natural antimicrobial agents.
- **Improved Formulations and Delivery Systems:** Innovations in the formulation of disinfectants and antiseptics, such as encapsulation techniques, can enhance their stability, release profiles, and efficacy. These advancements improve their performance in healthcare and other settings.
- **Non-Chemical Alternatives:** Non-chemical disinfection technologies, such as UV-C light and ozone treatment, are gaining traction as safe and eco-friendly options for sterilizing surfaces, water, and air.

18.7 CONCLUSION

Chemical disinfectants, antiseptics, and preservatives are indispensable for infection control and product preservation across various industries. Despite the challenges associated with their use, these agents continue to protect public health, prolong product shelf life, and prevent microbial contamination. As science progresses, innovations in these fields will likely lead to safer, more effective, and environmentally friendly solutions, meeting the growing demands of healthcare, food safety, and sustainability.

CHAPTER 19
LABORATORY EVALUATION OF ANTIMICROBIAL AGENTS

The evaluation of antimicrobial agents is a critical component of both clinical microbiology and pharmaceutical development, aimed at assessing the efficacy of substances against microbial pathogens. This process is essential for developing new drugs and ensuring that existing ones maintain their effectiveness. With the rise of antimicrobial resistance, there is a heightened need to identify agents that can combat resistant strains effectively. Laboratory evaluation provides insights into the potency, spectrum, and safety of these agents, guiding clinical applications and informing prescribing practices.

19.1 METHODS FOR ASSESSING ANTIMICROBIAL ACTIVITY

Laboratory methods for evaluating antimicrobial agents fall into two primary categories: in vitro and in vivo techniques. In vitro methods, conducted in a controlled environment outside a living organism, allow for a more rapid and controlled assessment of antimicrobial properties. In vivo methods, by contrast, involve testing within a living organism, such as in animal models, providing insights into pharmacodynamics and pharmacokinetics.

19.1.1 Disc Diffusion Method

The disc diffusion method, or Kirby-Bauer test, is a commonly used qualitative method to evaluate the efficacy of an antimicrobial agent. In this method, discs impregnated with the antimicrobial agent are placed on an agar plate inoculated with the test microorganism. As the agent diffuses into the agar, it inhibits bacterial growth around the disc if it is effective against the microorganism. This creates a clear zone, termed the "zone of inhibition," the size of which reflects the potency of the antimicrobial agent. Although the disc diffusion method is widely used due to its simplicity, it does not provide quantitative data on the minimum inhibitory concentration (MIC).

19.1.2 Minimum Inhibitory Concentration (MIC) Testing

MIC testing is a quantitative method that measures the lowest concentration of an antimicrobial agent required to inhibit visible growth of a microorganism. Typically, the MIC is determined using a broth microdilution method or an E-test. In the broth microdilution method, various concentrations of the antimicrobial agent are prepared in a series of tubes or microplate wells, and each is inoculated with a standard amount of the microorganism. The MIC is the lowest concentration where no visible growth is observed. This technique is invaluable for determining effective dosages and understanding the potency of new or existing antimicrobial agents.

19.1.3 Minimum Bactericidal Concentration (MBC) Testing

MBC testing determines the lowest concentration of an antimicrobial agent required to kill a specific microorganism. This method involves subculturing broth dilutions from an MIC test that show no visible growth onto drug-free agar plates. The MBC is defined as the lowest concentration of the antimicrobial that produces no growth on the agar plate, meaning the agent has killed the organism. MBC testing is particularly useful for distinguishing between bactericidal agents, which kill bacteria, and bacteriostatic agents, which only inhibit bacterial growth.

19.1.4 Time-Kill Assay

The time-kill assay is a dynamic method used to evaluate the rate at which an antimicrobial agent kills a microorganism over time. In this test, the organism is exposed to a fixed concentration of the antimicrobial, and samples are taken at various time intervals to measure bacterial counts. Time-kill assays provide valuable information about the kinetics of microbial killing, helping determine how quickly an agent can reduce bacterial loads and at what concentrations. Such data are especially useful in understanding the drug's efficacy in real-world treatment settings.

19.2 EVALUATING SPECTRUM OF ACTIVITY

One important aspect of antimicrobial evaluation is determining the spectrum of activity, which describes the range of pathogens the agent can effectively combat. Antimicrobials may be classified as broad-spectrum or narrow-spectrum based on their effectiveness across different types of bacteria, including gram-positive and gram-negative strains.

19.2.1 Broad-Spectrum Antimicrobials

Broad-spectrum agents are effective against a wide variety of microbial species. These are especially valuable in situations where the pathogen is unknown, as they can target multiple possible causative agents simultaneously. However, broad-spectrum use increases the risk of disrupting normal microbiota and can contribute to resistance if not used judiciously.

19.2.2 Narrow-Spectrum Antimicrobials

Narrow-spectrum antimicrobials target specific types or groups of microorganisms, such as gram-positive bacteria only. These agents are preferable when the causative organism is known, as they are less likely to disturb the microbiota and contribute to resistance. Laboratory testing aids in identifying the specific spectrum of an antimicrobial agent, allowing for more targeted therapeutic use.

19.3 ASSESSING ANTIMICROBIAL RESISTANCE

The rise of antimicrobial resistance (AMR) has prompted the need for robust laboratory methods to monitor and evaluate resistance mechanisms. AMR testing is vital in identifying resistance patterns in pathogens, which informs clinical decision-making and infection control practices.

19.3.1 Antibiotic Susceptibility Testing (AST)

AST involves evaluating the susceptibility of pathogens to different antimicrobial agents. It encompasses methods like disc diffusion, MIC determination, and automated systems. The results of AST help in choosing effective treatments and in monitoring resistance trends within populations over time.

19.3.2 Genotypic Methods for Detecting Resistance

Genotypic testing identifies specific resistance genes within microorganisms. Techniques like polymerase chain reaction (PCR) and sequencing can detect known resistance markers, helping in understanding the genetic basis of resistance. These methods are particularly useful for identifying novel resistance mechanisms and for tracking the spread of resistant strains.

19.3.3 Phenotypic Resistance Testing

Phenotypic testing involves observing the growth patterns of bacteria in the presence of an antimicrobial agent to identify resistance phenotypes. This includes tests like the β-lactamase test, which identifies organisms that produce enzymes capable of breaking down β-lactam antibiotics. Phenotypic resistance testing complements genotypic methods by confirming resistance profiles in clinical isolates.

19.4 EVALUATING SAFETY AND TOXICITY

An essential component of antimicrobial evaluation involves assessing the safety and toxicity of agents to ensure they are safe for human use. Laboratory studies initially test for cytotoxicity and potential adverse effects in cell cultures and animal models.

19.4.1 Cytotoxicity Assays

Cytotoxicity assays evaluate the toxic effects of antimicrobial agents on mammalian cells. These tests determine concentrations at which the drug is safe for human cells, which is critical in defining therapeutic doses and reducing side effects.

19.4.2 Animal Studies

Animal models are used to study the pharmacokinetics and pharmacodynamics of antimicrobial agents in vivo. These studies help researchers understand how the drug behaves within a living organism, including its distribution, metabolism, and excretion. Animal studies provide crucial data on safety, potential side effects, and dosage levels, which is fundamental in moving a drug to clinical trials.

19.5 IMPORTANCE OF QUALITY CONTROL IN ANTIMICROBIAL TESTING

Quality control (QC) measures are vital in laboratory evaluation to ensure that tests produce accurate and reproducible results. In antimicrobial testing, QC involves using control strains with known susceptibility profiles to validate test procedures and detect any inconsistencies. Regulatory bodies, such as the Clinical and Laboratory Standards Institute (CLSI) and the European Committee on Antimicrobial Susceptibility Testing (EUCAST), provide guidelines and reference standards for performing antimicrobial tests. Adhering to these standards ensures reliable results that can be compared across studies and institutions.

19.6 CONCLUSION

The laboratory evaluation of antimicrobial agents is a complex process that integrates a variety of testing methods to understand the efficacy, spectrum, resistance potential, and safety of these agents. Through in vitro and in vivo testing, researchers can identify potent antimicrobial agents, assess their appropriate usage, and track resistance trends. In an era marked by increasing antimicrobial resistance, these laboratory evaluations are vital for developing effective treatments and preserving the efficacy of existing antimicrobial agents.

CHAPTER 20

MICROBICIDES: MODE OF ACTION AND RESISTANCE

Microbicides are substances specifically designed to reduce the transmission of sexually transmitted infections (STIs) by acting as protective agents when applied to the vaginal or rectal mucosa. They represent a promising approach to controlling the spread of various pathogens, including HIV, herpes simplex virus (HSV), and human papillomavirus (HPV). With the development of new and innovative formulations, microbicides offer potential as a preventive tool, especially in resource-limited settings or where other preventive options like condoms are not always feasible or accessible.

20.1 MODE OF ACTION OF MICROBICIDES

Microbicides work through various mechanisms to inhibit or neutralize pathogens at different stages of infection. Their action is primarily localized to the site of application, and they target a broad spectrum of pathogens, ranging from viruses to bacteria and fungi. Some microbicides aim to prevent infection by blocking the pathogen from binding to host cells, while others act by directly inactivating the pathogens. The mode of action of a microbicide generally depends on its chemical nature and the specific pathogen it targets.

20.1.1 Inhibition of Pathogen Binding

Many microbicides work by preventing pathogens from attaching to the host cells, a critical initial step in infection. Certain compounds inactivate cell surface receptors or block viral proteins that facilitate attachment to mucosal cells. For instance, some HIV-targeting microbicides contain entry inhibitors, which prevent the virus from binding to CD4 receptors on T cells. This approach effectively reduces the likelihood of pathogen entry and replication in host cells, limiting the opportunity for infection to establish.

20.1.2 Disruption of Pathogen Membranes

Some microbicides exert their effect by disrupting the structural integrity of pathogens. Agents such as surfactants destabilize the viral envelope or bacterial membrane, rendering the pathogens incapable of functioning or replicating. For instance, nonoxynol-9 (N-9) is a surfactant-based microbicide that disrupts the lipid membranes of pathogens, effectively killing them. However, the use of surfactants like N-9 has also been associated with mucosal irritation, potentially increasing susceptibility to infection, which underscores the importance of safety in microbicide development.

20.1.3 Interference with Viral Replication

Certain microbicides specifically target viral replication mechanisms. Reverse transcriptase inhibitors, for example, interfere with the ability of HIV to transcribe its RNA into DNA, a necessary step for viral replication. By halting this process, these microbicides can effectively limit the spread of the virus within the body. Microbicides that inhibit replication are particularly effective because they target processes that are essential for the virus to proliferate, rather than just preventing the initial infection.

20.1.4 Enhancement of Host Defense Mechanisms

Some microbicides are designed to boost the body's natural defense systems. These agents stimulate local immune responses, helping the host body to fight off infections more effectively. For example, certain microbicides induce the production of protective mucosal antibodies or increase the activity of immune cells within the genital tract. This approach not only provides immediate protection but may also offer longer-lasting immunity by priming the immune system to recognize and respond to specific pathogens.

20.2 CLASSES OF MICROBICIDES BASED ON CHEMICAL COMPOSITION

Microbicides can be classified based on their chemical composition and mechanisms of action. Each class is designed with specific purposes, benefits, and limitations that influence their effectiveness against different pathogens.

20.2.1 Surfactant-Based Microbicides

Surfactants are among the earliest forms of microbicides, developed for their ability to disrupt the lipid membranes of pathogens. However, as mentioned earlier, surfactant-based microbicides can be irritating to mucosal tissues, potentially making users more susceptible to infection. Although effective against a range of pathogens, the safety concerns have limited their widespread use.

20.2.2 Polyanion-Based Microbicides

Polyanions, such as sulfated polysaccharides, work by binding to positively charged proteins on the surfaces of pathogens, effectively preventing them from attaching to host cells. Polyanion microbicides, including agents like cellulose sulfate, show promise against multiple viruses but have shown mixed results in clinical trials.

Further research aims to enhance their efficacy and safety profile.

20.2.3 Antiretroviral-Based Microbicides

Antiretroviral (ARV)-based microbicides, which contain drugs commonly used in HIV treatment, have shown high efficacy in preventing HIV transmission. Examples include tenofovir-based microbicides, which inhibit HIV replication and have been tested successfully in gel formulations. These microbicides offer a more targeted approach, as they specifically interfere with the viral life cycle, making them especially useful for high-risk populations.

20.2.4 Protein-Based Microbicides

Protein-based microbicides, including lectins and monoclonal antibodies, are a newer category of microbicides that work by binding directly to viral proteins. This binding blocks the virus from interacting with host cells, effectively neutralizing it. Although promising, protein-based microbicides are still in early development stages, and their cost and stability issues present challenges for broader use.

20.3 DEVELOPMENT CHALLENGES AND SAFETY CONSIDERATIONS

While microbicides hold great promise as preventive agents, their development faces significant challenges. Ensuring safety, efficacy, and user compliance are crucial for their success. Some microbicides, like surfactants, have shown adverse effects, such as irritation of mucosal tissues, which could increase susceptibility to infections. Additionally, optimizing formulations to provide consistent dosing and stability in varying conditions is essential. Delivery methods, such as gels, rings, and films, are also under investigation to maximize ease of use and effectiveness.

User adherence remains another critical challenge, as the effectiveness of microbicides often depends on regular application. Long-acting delivery systems, like vaginal rings, are being explored to address this issue, as they require less frequent application and can deliver consistent doses over time.

20.4 MECHANISMS OF RESISTANCE TO MICROBICIDES

Microbicides, like antibiotics and antiviral drugs, face the challenge of microbial resistance. The development of resistance to microbicides is influenced by various factors, including improper use, prolonged exposure, and suboptimal concentrations. Microbial resistance reduces the efficacy of microbicides and may limit their usefulness as preventive agents over time.

20.4.1 Genetic Mutations

Some pathogens, especially viruses like HIV, can mutate rapidly, leading to resistance against specific microbicides. For example, HIV may develop mutations in the gene coding for reverse transcriptase, rendering certain ARV-based microbicides ineffective. These mutations can arise in response to prolonged exposure to low concentrations of the microbicide, highlighting the importance of correct dosing and administration.

20.4.2 Biofilm Formation in Bacteria

Bacterial pathogens can form biofilms, which are clusters of bacteria encased in a protective matrix. This matrix shields the bacteria from antimicrobial agents, allowing them to survive even in the presence of microbicides. Biofilms can complicate the treatment of bacterial STIs and reduce the effectiveness of microbicidal agents, as they hinder penetration and distribution of the active compounds.

20.4.3 Efflux Pumps and Other Resistance Mechanisms

Some bacteria possess efflux pumps, which actively expel antimicrobial agents from their cells. Efflux pumps allow pathogens to resist the effects of microbicides by limiting the intracellular concentration of these agents. Certain strains of bacteria have also developed enzymatic mechanisms to inactivate microbicidal agents, further complicating treatment.

20.5 STRATEGIES TO OVERCOME RESISTANCE

Overcoming resistance to microbicides requires a multifaceted approach, involving combination therapies, targeted formulations, and new drug development. By combining microbicides with different mechanisms of action, researchers aim to reduce the likelihood of resistance development. For example, using both entry inhibitors and replication inhibitors could limit the ability of pathogens to adapt and survive. Developing microbicides with improved stability and delivery methods is also essential to ensure consistent and effective dosing. Advanced delivery systems, such as nanotechnology-based formulations, can enhance the penetration of microbicides into biofilms or directly target specific pathogens. Ongoing research into alternative chemical structures and novel compounds may provide the next generation of microbicides with broader spectra of activity and improved resistance profiles.

20.6 CONCLUSION

Microbicides offer a valuable tool in preventing the spread of sexually transmitted infections, providing a critical option for individuals seeking additional protection against pathogens. Their effectiveness depends on understanding and leveraging their modes of action while addressing challenges related to resistance. As new formulations and delivery methods are developed, microbicides may become an increasingly reliable and accessible preventive measure.

CHAPTER 21

STERILIZATION PROCEDURES AND STERILITY ASSURANCE

Sterilization is a fundamental process in healthcare, research, pharmaceuticals, and food production, aimed at eliminating all forms of microbial life, including bacteria, viruses, fungi, and spores. The primary goal of sterilization is to ensure that products and environments are free from any microorganisms that could potentially cause contamination, infection, or spoilage. Sterility assurance, meanwhile, involves a systematic approach to verifying and validating sterilization procedures, ensuring that they consistently meet defined standards. Together, sterilization procedures and sterility assurance play a critical role in maintaining high standards of safety and efficacy across various industries.

21.1 TYPES OF STERILIZATION PROCEDURES

Sterilization methods vary based on the type of item, material, or environment being sterilized and the specific microorganisms targeted. The main sterilization procedures include heat sterilization, chemical sterilization, filtration, and radiation.

21.1.1 Heat Sterilization

Heat sterilization is one of the most common methods used, relying on high temperatures to kill microorganisms. It can be further divided into two main types: moist heat and dry heat sterilization.

- **Moist Heat Sterilization:** This method uses steam under pressure to achieve sterilization. Autoclaving, a common technique in healthcare and laboratory settings, involves exposing items to steam at 121°C to 134°C under pressure. Moist heat sterilization is effective against a wide range of pathogens, including bacterial spores, which are highly resistant to other forms of sterilization.
- **Dry Heat Sterilization:** In dry heat sterilization, items are exposed to hot air at temperatures ranging from 160°C to 180°C for extended periods. Unlike moist heat, dry heat does not involve pressure and is generally used for materials that cannot tolerate moisture, such as oils, powders, and glassware.

21.1.2 Chemical Sterilization

Chemical sterilization involves using chemical agents to achieve sterility, making it suitable for heat-sensitive items. Common chemical sterilants include ethylene oxide gas, hydrogen peroxide vapor, and formaldehyde. These agents disrupt microbial cell walls or interfere with cellular processes, leading to cell death. Ethylene oxide sterilization is frequently used in the medical device industry due to its effectiveness in penetrating complex devices with hard-to-reach surfaces. However, it requires careful handling due to its toxicity and potential environmental impact.

21.1.3 Filtration

Filtration sterilization is used mainly for liquids and gases that are sensitive to heat and chemicals. By passing the substance through a membrane filter with pores smaller than the size of microorganisms, it is possible to remove contaminants. Filtration is a common method for sterilizing pharmaceuticals, culture media, and injectable solutions. However, filtration does not eliminate all viruses or prions, and as such, it may require combination with other sterilization methods depending on the application.

21.1.4 Radiation Sterilization

Radiation sterilization uses ionizing radiation to kill microorganisms. Two main types of radiation used are gamma radiation and electron beam (e-beam) sterilization. Gamma radiation, commonly applied in medical device sterilization, penetrates deeply and is highly effective, while e-beam sterilization is faster and suitable for less dense materials. Radiation sterilization is advantageous for single-use medical products, but it requires specific equipment and regulatory considerations.

21.2 STERILITY ASSURANCE AND QUALITY CONTROL

Sterility assurance is a systematic approach that ensures sterilization processes consistently achieve the desired level of microbial elimination. The sterility assurance level (SAL) is a measure of the probability of a single microorganism surviving after sterilization, typically set at 10^-6, meaning there is less than one chance in a million of a microorganism surviving.

21.2.1 Validation of Sterilization Processes

Validation is an essential part of sterility assurance, confirming that a sterilization process consistently meets requirements. This includes testing the sterilization cycle parameters (e.g., time, temperature, chemical concentration), as well as verifying the efficacy of equipment used in sterilization. Validation also involves conducting microbiological testing to ensure that microbial reduction goals are met.

21.2.2 Biological Indicators

Biological indicators (BIs) are used in sterility assurance as a standard for evaluating the effectiveness of sterilization processes. BIs contain known concentrations of highly resistant bacterial spores, such as *Geobacillus stearothermophilus* (used for moist heat sterilization) or *Bacillus atrophaeus* (used for ethylene oxide and dry heat sterilization). Following sterilization, BIs are cultured to check for microbial growth. A lack of growth indicates successful sterilization, as even the most resistant organisms have been eliminated.

21.2.3 Chemical Indicators and Integrators

Chemical indicators are often used as a quick visual check to confirm that sterilization parameters, such as temperature and exposure time, were reached. These indicators change color or form when exposed to specific conditions. Chemical integrators are more sophisticated, designed to respond to multiple parameters of sterilization (e.g., time, temperature, and moisture), providing more reliable assurance compared to single-parameter indicators. While they are not replacements for BIs, chemical indicators play an essential role in routine quality control.

21.2.4 Monitoring and Documentation

Effective monitoring and documentation are essential components of sterility assurance. This includes recording the parameters of each sterilization cycle, tracking the performance of equipment, and documenting any deviations from standard protocols. Comprehensive records allow for quick identification of issues, facilitate troubleshooting, and provide evidence of compliance with regulatory standards.

21.3 REGULATORY REQUIREMENTS AND COMPLIANCE

Compliance with regulatory standards is crucial in sterility assurance, especially in healthcare, pharmaceuticals, and food industries. Various regulatory bodies, including the U.S. Food and Drug Administration (FDA), the International Organization for Standardization (ISO), and the World Health Organization (WHO), set guidelines and requirements to ensure that sterilization processes are safe, effective, and consistently reliable.

21.3.1 FDA and ISO Standards

The FDA oversees sterilization processes in the medical and pharmaceutical industries in the United States, enforcing standards that ensure patient safety and product efficacy. The ISO has developed international standards, such as ISO 11135 for ethylene oxide sterilization and ISO 11137 for radiation sterilization, which outline requirements for process validation, testing, and quality control. These standards help establish uniform practices across industries and regions, fostering global reliability in sterilization processes.

21.3.2 Good Manufacturing Practices (GMP)

Good Manufacturing Practices (GMP) are guidelines that address every aspect of production, from facility design to employee hygiene and equipment maintenance. GMP emphasizes quality assurance and sterility at every stage of production, and it is mandatory in regulated industries, such as pharmaceuticals and medical devices. Compliance with GMP helps prevent contamination and ensures that each batch of products meets quality and safety requirements.

21.4 CHALLENGES IN STERILIZATION AND STERILITY ASSURANCE

Despite advances in sterilization techniques, challenges remain, particularly when dealing with complex products and resistant microorganisms.

21.4.1 Sterilizing Complex Medical Devices

Medical devices with intricate designs, such as catheters, endoscopes, and surgical instruments, are difficult to sterilize effectively.

These devices often contain narrow lumens or enclosed areas that can harbor microbes, making it challenging for sterilizing agents to reach all surfaces. Manufacturers must design devices with sterilization in mind, and sterilization procedures must be validated carefully to ensure that all areas are adequately treated.

21.4.2 Emergence of Resistant Microorganisms

Some microorganisms, such as bacterial spores, prions, and biofilms, exhibit high resistance to standard sterilization methods. Biofilms, in particular, are problematic as they form protective layers that shield microbes from chemical agents and radiation. Prions, the infectious agents responsible for conditions like Creutzfeldt-Jakob disease, are highly resistant to conventional sterilization, requiring specialized processes. Addressing these resistant forms remains a critical focus in research and process development.

21.4.3 Balancing Efficacy and Material Compatibility

Certain sterilization methods, like high heat or aggressive chemicals, can damage sensitive materials, limiting the choice of sterilization techniques for specific products. For example, some plastic-based medical devices may become brittle under heat or chemical exposure. Sterilization processes must strike a balance between effective microbial eradication and preserving the integrity of the materials being sterilized.

21.5 CONCLUSION

Sterilization and sterility assurance are foundational components of safety in various sectors, from healthcare to pharmaceuticals and food production. The development and validation of reliable sterilization processes ensure that products meet the highest standards of cleanliness and safety, protecting consumers and patients from potential harm.

As technology and understanding of microbial resistance evolve, new methods and enhanced sterility assurance practices will be needed to address the complexities and challenges in this critical area. By continually advancing sterilization science, industries can better safeguard public health and meet stringent regulatory requirements.

PART 6:

PHARMACEUTICAL PRODUCTION

CHAPTER 22

STERILE PHARMACEUTICAL PRODUCTS AND PRINCIPLES OF GOOD MANUFACTURING PRACTICE

Sterile pharmaceutical products, such as injectable drugs, ophthalmic solutions, and certain creams and ointments, must meet the highest standards of cleanliness and safety to ensure they are free of contaminants and safe for human use. Sterility in pharmaceuticals is crucial because these products are introduced directly into the body, often bypassing natural barriers that typically filter out contaminants. As such, they pose a high risk of infection if contaminated with microorganisms or particles.

To minimize these risks, the production of sterile pharmaceuticals requires strict adherence to Good Manufacturing Practice (GMP) principles, which encompass a comprehensive set of guidelines designed to ensure product quality, safety, and consistency.

22.1 IMPORTANCE OF STERILITY IN PHARMACEUTICALS

Sterility is essential in pharmaceuticals to protect patients from potentially life-threatening infections. Any contamination in sterile pharmaceuticals can result in severe adverse reactions, including sepsis and organ failure. Sterile pharmaceutical products are often used by vulnerable populations, including immunocompromised individuals and patients in critical care settings, where even minor contamination can have serious consequences. For this reason, sterility in these products is non-negotiable, and manufacturers are required to implement stringent protocols that eliminate the risk of contamination.

22.2 TYPES OF STERILE PHARMACEUTICAL PRODUCTS

22.2.1 Injectables

Injectable drugs are among the most common sterile products, delivered directly into the bloodstream or tissues. Since they bypass the body's natural defenses, such as the digestive tract, they require absolute sterility. Injectable pharmaceuticals include intravenous (IV), intramuscular (IM), and subcutaneous (SC) formulations, each with specific requirements for sterility, formulation, and packaging.

22.2.2 Ophthalmic Solutions

Ophthalmic products, such as eye drops and eye ointments, are also classified as sterile due to the sensitivity of the eyes and the risk of infection. Contaminants in ophthalmic products can cause infections, irritation, or even blindness, making sterility a vital requirement.

22.2.3 Inhalation Products

Inhalable pharmaceuticals, such as those used in nebulizers and inhalers, need to be sterile to prevent respiratory infections. The lungs are particularly vulnerable to microbial contamination due to their direct exposure to the bloodstream and the large surface area available for absorption.

22.2.4 Topical Preparations and Implants

Certain creams, ointments, and implants that penetrate the skin or mucous membranes also require sterility, particularly when used on open wounds or in surgical settings. These products must be free from bacteria, fungi, and other pathogens to prevent infections and support healing.

22.3 PRINCIPLES OF GOOD MANUFACTURING PRACTICE (GMP) FOR STERILE PHARMACEUTICALS

Good Manufacturing Practice (GMP) refers to a set of regulatory guidelines intended to ensure consistent quality and safety in pharmaceutical production. GMP encompasses every aspect of production, from facility design and equipment maintenance to employee training and quality control. The principles of GMP are particularly stringent for sterile products due to their high risk and sensitivity.

22.3.1 Facility Design and Environmental Control

The design of the facility plays a significant role in maintaining sterility. GMP guidelines mandate that sterile manufacturing facilities must have controlled environments with cleanrooms, air filtration, and pressure differentials to prevent contamination. Cleanrooms are classified according to the number of allowable particles per cubic meter, with stricter limits for sterile products. For example, aseptic areas where products are exposed to the environment should be ISO Class 5 (Grade A), with minimal particle and microbial contamination. Maintaining controlled humidity and temperature levels also helps in minimizing microbial growth and ensuring product stability.

22.3.2 Personnel Training and Hygiene

Personnel involved in sterile manufacturing must undergo rigorous training in aseptic techniques, gowning procedures, and contamination control. Proper gowning is essential to prevent contamination from skin, hair, or other particles shed by individuals. Workers in critical areas must wear sterile gowns, gloves, masks, and other protective gear. Furthermore, GMP emphasizes regular monitoring of personnel to ensure compliance with hygiene protocols and minimize the risk of contamination in sensitive environments.

22.3.3 Equipment Sterilization and Maintenance

All equipment used in the production of sterile pharmaceuticals must be sterilized before use to avoid introducing contaminants. This includes autoclaves, filtration systems, and filling machines.

Equipment sterilization can be achieved through methods such as heat sterilization, chemical sterilization, or irradiation, depending on the nature of the equipment and the product. Regular maintenance and validation are also crucial to ensure that equipment operates correctly and continues to meet sterility standards.

22.3.4 Raw Material Handling and Quality Control

GMP guidelines require strict control over raw materials to ensure they meet defined standards of purity and quality.

Contaminated raw materials are a significant risk factor in pharmaceutical production, as they can introduce microorganisms or particles into the final product. Incoming raw materials must be tested and stored under conditions that prevent microbial growth or degradation. Additionally, raw materials should be handled in dedicated areas or facilities to prevent cross-contamination.

22.3.5 Aseptic Processing and Filling

Aseptic processing is critical in sterile pharmaceutical manufacturing, especially when the final product cannot be terminally sterilized. Aseptic filling involves transferring the product into its final container in a controlled environment to maintain sterility.

Equipment and containers are sterilized beforehand, and the entire process occurs in a Class 100 (ISO 5) environment. GMP standards mandate rigorous monitoring of environmental conditions during aseptic filling, with strict protocols for handling containers, seals, and caps to prevent contamination.

22.3.6 Packaging and Labeling

Packaging plays an essential role in maintaining product sterility throughout its shelf life. Containers, such as vials, ampoules, and syringes, must be designed to withstand sterilization methods and maintain an airtight seal. GMP also requires that labels be clear and include important information, such as dosage instructions, batch numbers, and expiration dates. Proper labeling helps ensure traceability, quality control, and adherence to regulatory standards.

22.4 QUALITY CONTROL AND STERILITY ASSURANCE

Ensuring the sterility of pharmaceutical products requires continuous quality control and testing. GMP guidelines establish specific sterility assurance practices to verify that all batches meet required standards.

22.4.1 Sterility Testing

Sterility testing is performed on each batch of sterile products to confirm the absence of viable microorganisms. Testing involves incubating samples in nutrient-rich media to encourage microbial growth if any are present. Products are tested under controlled conditions, and results are monitored for a specified period to detect any contamination.

22.4.2 Endotoxin Testing

Endotoxins are toxic substances produced by certain bacteria, primarily Gram-negative bacteria. Even if the bacteria are killed during sterilization, endotoxins may remain and cause harmful reactions, such as fever and shock, in patients. Endotoxin testing is essential for sterile products, especially injectables, to ensure that no endotoxins are present in the final product.

22.4.3 Environmental Monitoring

Continuous environmental monitoring is an integral part of GMP for sterile manufacturing facilities. This includes air quality testing for particles and microorganisms, as well as surface and personnel monitoring. Environmental monitoring helps detect potential sources of contamination early, allowing manufacturers to take corrective action before issues impact product quality.

22.4.4 Process Validation

Validation is critical in GMP, ensuring that each step in the manufacturing process reliably achieves the intended outcomes. Process validation confirms that sterilization methods, such as autoclaving or filtration, consistently remove or kill contaminants to the required sterility assurance level. Validation is also applied to aseptic processing techniques to verify that they effectively prevent contamination.

22.5 REGULATORY COMPLIANCE IN STERILE MANUFACTURING

Compliance with regulatory agencies, such as the U.S. Food and Drug Administration (FDA), the European Medicines Agency (EMA), and the World Health Organization (WHO), is vital for pharmaceutical manufacturers.

These agencies set strict GMP standards for sterile products, requiring companies to maintain detailed records, conduct regular inspections, and implement quality management systems. Compliance with GMP ensures that companies produce safe, effective, and reliable sterile products that meet the requirements of global markets.

22.6 CHALLENGES IN STERILE PHARMACEUTICAL PRODUCTION

Producing sterile pharmaceuticals presents unique challenges, particularly with complex formulations and advanced therapeutic agents like biologics. Biologic drugs, for example, are highly sensitive and require specialized handling to maintain sterility without compromising efficacy. Additionally, the rise in multidrug-resistant organisms has led to stricter standards and a greater emphasis on preventing contamination from the environment, personnel, and equipment. Meeting these challenges requires ongoing investment in advanced technology, robust quality control systems, and personnel training.

22.7 CONCLUSION

The production of sterile pharmaceutical products is a highly regulated process that requires meticulous attention to detail and adherence to GMP principles. By implementing stringent quality control and sterility assurance measures, pharmaceutical manufacturers can produce safe and effective sterile products that protect patient health. Continuous improvement, investment in technology, and regulatory compliance are essential in meeting the high standards required for sterile pharmaceuticals, helping to ensure safety and quality for patients worldwide.

CHAPTER 23

THE MANUFACTURE AND QUALITY CONTROL OF IMMUNOLOGICAL PRODUCTS

Immunological products, including vaccines, monoclonal antibodies, and immune globulins, play an essential role in the prevention, treatment, and management of various infectious diseases, autoimmune disorders, and cancers.

These products work by either stimulating the immune system to fight against specific pathogens or modulating immune responses in certain conditions. Given their direct impact on human health, the manufacture of immunological products demands rigorous quality control measures to ensure safety, efficacy, and consistency. These measures are governed by Good Manufacturing Practice (GMP) standards, which outline best practices for each stage of production.

23.1 TYPES OF IMMUNOLOGICAL PRODUCTS

23.1.1 Vaccines

Vaccines are designed to stimulate the immune system to recognize and combat specific pathogens, such as viruses or bacteria.

They often contain weakened or inactivated forms of pathogens, or their specific proteins, which train the immune system without causing disease. Vaccines are used to prevent infectious diseases, making their safety and efficacy critical in public health.

23.1.2 Monoclonal Antibodies

Monoclonal antibodies (mAbs) are laboratory-made molecules that can bind to specific antigens, such as cancer cells or viral proteins. These antibodies are tailored to target specific cells, marking them for destruction by the immune system or directly inhibiting their activity. Monoclonal antibodies are used in treating cancer, autoimmune diseases, and infectious diseases, among others.

23.1.3 Immune Globulins

Immune globulins are preparations made from pooled human plasma, containing antibodies against various pathogens. They are used to provide passive immunity to individuals who may be at high risk of infection or who have weakened immune systems. Immune globulins are essential for providing temporary immunity and are commonly used in immunocompromised individuals.

23.2 MANUFACTURING PROCESS OF IMMUNOLOGICAL PRODUCTS

The production of immunological products is complex and must adhere to strict guidelines to ensure their safety and effectiveness. The process varies depending on the type of product but generally includes several critical stages.

23.2.1 Antigen Preparation

For vaccines, the first step is the preparation of antigens, which are components of pathogens that stimulate an immune response. This can involve growing viruses or bacteria in controlled environments or using genetic engineering to produce recombinant proteins. The method of antigen production varies depending on whether the vaccine is live, inactivated, or subunit-based.

23.2.2 Culturing and Fermentation

Monoclonal antibodies and some vaccines require culturing cells in bioreactors to produce the necessary proteins or antigens. During the fermentation process, cell cultures are grown in sterile environments under controlled conditions, including temperature, pH, and nutrient supply. This process ensures that high yields of the desired antigen or antibody are obtained.

23.2.3 Purification and Concentration

Once the antigen or antibody is produced, it must be purified to remove any impurities, such as unwanted proteins, cell debris, or endotoxins. Techniques like chromatography, filtration, and centrifugation are used to achieve the required purity levels. This step is crucial, as impurities can cause adverse reactions and affect the product's efficacy.

23.2.4 Inactivation and Stabilization

For inactivated vaccines, pathogens are treated with chemicals or heat to render them non-infectious while preserving their immunogenic properties. Stabilizers are also added to immunological products to maintain their effectiveness during storage and transport. These stabilizers ensure that the product remains potent over its intended shelf life.

23.2.5 Formulation and Fill-Finish

After purification and stabilization, the product is formulated with other components, such as adjuvants or preservatives, to enhance its efficacy and stability. The formulated product is then filled into sterile containers, such as vials or syringes, under aseptic conditions. The fill-finish process requires strict adherence to GMP standards to avoid contamination.

23.3 QUALITY CONTROL IN IMMUNOLOGICAL PRODUCT MANUFACTURING

Quality control (QC) is a critical aspect of immunological product manufacturing. It ensures that each batch of product meets the required safety, potency, and purity standards before being released to the market.

23.3.1 Sterility Testing

Sterility is crucial for immunological products to prevent infections in recipients. Each batch is tested for sterility using methods like direct inoculation or membrane filtration. These tests verify that the product is free from bacterial, fungal, or other microbial contamination. Sterility testing is particularly important for injectable immunological products.

23.3.2 Potency Testing

Potency testing is essential to confirm that the product will elicit the desired immune response. For vaccines, this involves determining the concentration of the antigen, while for monoclonal antibodies, it may involve testing the product's binding affinity to the target antigen. Potency tests are carefully standardized to ensure consistency across batches.

23.3.3 Purity and Safety Testing

Purity tests assess the presence of unwanted substances, including host cell proteins, DNA, or other impurities. Safety testing involves assays to detect endotoxins, which can cause severe reactions if present. Purity and safety testing are performed using techniques like high-performance liquid chromatography (HPLC) and gel electrophoresis to ensure that the product is free from harmful contaminants.

23.3.4 Stability Testing

Stability testing evaluates how well an immunological product retains its quality and efficacy over time under different storage conditions. This testing is crucial for determining the product's shelf life and optimal storage conditions. Stability tests simulate various environmental factors, such as temperature, humidity, and light exposure, to assess the product's robustness and longevity.

23.3.5 Identity and Consistency Testing

Identity testing ensures that the correct antigen or antibody is present in the final product. Consistency testing verifies that each batch meets the same quality standards as previous batches. This testing is vital for maintaining product uniformity, which is especially important in large-scale vaccine production.

23.4 REGULATORY COMPLIANCE IN IMMUNOLOGICAL PRODUCT MANUFACTURING

Manufacturing immunological products involves meeting stringent regulatory requirements set forth by organizations such as the U.S. Food and Drug Administration (FDA), European Medicines Agency (EMA), and World Health Organization (WHO). These regulations ensure that products are safe, effective, and produced to consistent standards.

23.4.1 Good Manufacturing Practice (GMP) Compliance

GMP regulations cover every aspect of production, from facility design and equipment to employee training and documentation. GMP compliance ensures that the product is consistently manufactured and controlled according to quality standards. Regular audits and inspections by regulatory agencies ensure that manufacturers adhere to GMP requirements.

23.4.2 Batch Release and Certification

Each batch of immunological products must be tested and certified before release. This involves extensive documentation, including batch records, test results, and quality control checks. Regulatory agencies may also require independent testing and certification of batches, particularly for vaccines, before they are distributed to the market.

23.4.3 Documentation and Traceability

Detailed documentation is required at every stage of production, from raw materials to final product testing. Traceability is crucial for identifying and addressing issues in specific batches, especially during recalls. Regulatory agencies require manufacturers to maintain these records for a specified period, ensuring accountability and transparency.

23.5 CHALLENGES IN IMMUNOLOGICAL PRODUCT MANUFACTURING

The production of immunological products presents unique challenges, largely due to the biological complexity of these products.

23.5.1 Biological Variability

Immunological products, especially those derived from living organisms, can exhibit variability between batches. This variability can affect product potency and efficacy, necessitating rigorous quality control and standardization to ensure consistency.

23.5.2 Contamination Risk

Due to their biological nature, immunological products are susceptible to contamination, which can compromise their safety and effectiveness. Maintaining aseptic conditions, using high-quality raw materials, and implementing stringent QC measures are essential for minimizing contamination risks.

23.5.3 Complex Regulatory Requirements

Immunological products must comply with a complex web of international regulations. This can create significant logistical and financial challenges for manufacturers, particularly in global distribution. Adhering to diverse regulatory standards requires substantial expertise and resources.

23.5.4 Cost and Production Scalability

The cost of manufacturing immunological products is often high, especially for advanced therapies like monoclonal antibodies. Scaling production to meet global demand without compromising quality is a constant challenge. Manufacturers must balance cost, scalability, and compliance to make these products widely accessible.

23.6 CONCLUSION

The manufacture and quality control of immunological products are complex processes that demand high standards to ensure product safety, efficacy, and consistency. Rigorous quality control measures, including sterility, potency, and stability testing, are essential to prevent contamination and ensure that each product meets regulatory standards. By adhering to Good Manufacturing Practices (GMP) and regulatory guidelines, manufacturers can produce reliable immunological products that play a critical role in protecting and enhancing human health worldwide. Advances in biotechnology and quality assurance processes continue to improve the quality and accessibility of immunological products, promising a brighter future in the fight against diseases.

CHAPTER 24
RECOMBINANT DNA TECHNOLOGY

Recombinant DNA technology, also known as genetic engineering, is a groundbreaking field of biotechnology that enables scientists to modify the genetic material of organisms. By combining DNA from different sources, this technology allows researchers to alter an organism's genes to create desirable traits, produce therapeutic proteins, or study gene function. Since its inception in the 1970s, recombinant DNA technology has revolutionized medicine, agriculture, and biological research, providing tools to create genetically modified organisms (GMOs), develop vaccines, and understand genetic diseases more thoroughly.

24.1 PRINCIPLES OF RECOMBINANT DNA TECHNOLOGY

The core principle of recombinant DNA technology lies in the ability to cut and reassemble DNA from different sources. This process typically involves several steps: isolation of target DNA, cutting DNA with restriction enzymes, inserting the target DNA into a vector, transferring the recombinant DNA into host cells, and selecting successful recombinants.

24.1.1 Isolation of Target DNA

The first step in recombinant DNA technology is to isolate the specific gene or DNA fragment that scientists wish to study or modify. This process often involves extracting DNA from the cells of an organism and purifying it to ensure that only the desired sequences are used in further steps. Techniques such as polymerase chain reaction (PCR) are also used to amplify specific DNA fragments, making them easier to work with.

24.1.2 Cutting DNA with Restriction Enzymes

Restriction enzymes, also known as molecular scissors, are proteins that recognize specific sequences of DNA and cut them at precise locations. This step is critical for creating DNA fragments that can be reassembled with other DNA segments. By cutting DNA in a controlled manner, restriction enzymes allow scientists to precisely manipulate genes, ensuring that the new combination of DNA is functional.

24.1.3 Inserting DNA into a Vector

After cutting the target DNA, scientists use vectors to insert the DNA fragment into host cells. Vectors are DNA molecules, like plasmids or viruses, that can replicate independently within a host cell. They act as carriers, ensuring that the recombinant DNA enters the cell and integrates into its genome. Vectors contain markers, like antibiotic resistance genes, which allow for the selection of successfully modified cells.

24.1.4 Transformation into Host Cells

The recombinant DNA is then introduced into host cells, typically bacteria, yeast, or mammalian cells, through a process called transformation. Transformation involves treating cells to make their membranes more permeable, allowing the DNA vector to enter. Once inside, the host cells begin to replicate and express the recombinant DNA, producing the desired protein or trait.

24.1.5 Selection and Screening

After transformation, researchers select the host cells that have successfully taken up the recombinant DNA. This is done using markers, such as antibiotic resistance, to identify transformed cells. Screening methods are then used to confirm that the recombinant DNA is correctly incorporated and functional.

24.2 APPLICATIONS OF RECOMBINANT DNA TECHNOLOGY

Recombinant DNA technology has diverse applications across fields like medicine, agriculture, and environmental science. Its versatility and precision make it invaluable for addressing pressing health, food security, and environmental challenges.

24.2.1 Medicine

Recombinant DNA technology has transformed medicine by enabling the production of proteins and hormones for therapeutic purposes. For example, human insulin, essential for diabetes management, was the first recombinant drug produced in the 1980s. Prior to this, insulin was extracted from animal sources, which posed challenges with purity and supply. Recombinant DNA technology also enables the production of growth hormones, blood clotting factors, and vaccines. Additionally, gene therapy, which involves modifying an individual's genes to treat genetic disorders, is a promising field driven by advancements in recombinant DNA technology.

24.2.2 Agriculture

In agriculture, recombinant DNA technology is used to create genetically modified crops that have improved resistance to pests, herbicides, and environmental stress. These genetically modified organisms (GMOs) increase crop yields, enhance nutritional value, and reduce the need for chemical pesticides. For example, Bt cotton and Bt corn are genetically engineered to produce a toxin derived from the bacterium *Bacillus thuringiensis*, which protects the crops from specific insect pests. Golden rice, engineered to contain higher levels of vitamin A, addresses nutrient deficiencies in regions with limited access to diverse diets.

24.2.3 Environmental Science

Recombinant DNA technology also has applications in environmental science, particularly in bioremediation, which is the use of organisms to clean up environmental pollutants. For example, bacteria can be genetically modified to break down oil spills or to degrade harmful chemicals in contaminated soils. Recombinant microorganisms can also be used in waste treatment and recycling processes, where they help to decompose waste products more efficiently.

24.2.4 Research and Development

In research, recombinant DNA technology is crucial for studying gene function and understanding the underlying mechanisms of diseases. By manipulating specific genes in model organisms, scientists can investigate their effects on biological processes. This knowledge is essential for developing new drugs, understanding genetic diseases, and advancing fields like developmental biology and neuroscience. Recombinant DNA technology also plays a key role in drug discovery, enabling scientists to produce proteins and enzymes needed to screen potential drug candidates.

24.3 ETHICAL AND SAFETY CONSIDERATIONS

While recombinant DNA technology offers significant benefits, it also raises ethical and safety concerns, particularly regarding the modification of living organisms and potential risks to human health and the environment.

24.3.1 Ethical Concerns

The use of genetically modified organisms, particularly in agriculture and food production, has led to debates over safety, labeling, and consumer choice. Some argue that GMOs may pose health risks, though extensive research has found no adverse effects linked to approved GMO products. Additionally, concerns about biodiversity arise with the widespread cultivation of GMO crops, as they may impact non-target organisms and reduce genetic diversity.

24.3.2 Biosafety and Containment

Recombinant DNA technology involves the use of bacteria and viruses as vectors, which could potentially spread genes unintentionally if not properly contained. Biosafety protocols are critical to prevent accidental release into the environment, particularly when working with pathogens or genes that confer antibiotic resistance. Laboratories handling recombinant DNA must follow containment guidelines set by organizations like the World Health Organization (WHO) and the Centers for Disease Control and Prevention (CDC).

24.3.3 Gene Editing and Human Germline Modifications

With the advent of gene-editing technologies like CRISPR, ethical debates around editing the human germline—changes that can be passed down to future generations—have intensified. While gene therapy can offer cures for genetic diseases, modifying the human genome in ways that could be inherited raises questions about long-term consequences and potential misuse. Regulatory frameworks are being developed globally to ensure that such technologies are used responsibly and ethically.

24.4 ADVANCES AND FUTURE DIRECTIONS IN RECOMBINANT DNA TECHNOLOGY

Recombinant DNA technology is rapidly evolving, with new tools and methods expanding its potential applications. CRISPR-Cas9, a precise gene-editing tool, has revolutionized genetic engineering by allowing scientists to make targeted changes to DNA with greater ease and accuracy. This advancement has opened up possibilities for correcting genetic mutations, studying gene function with more precision, and developing novel treatments for genetic disorders. Another emerging field is synthetic biology, which involves designing and constructing new biological parts, devices, and systems. By combining synthetic biology with recombinant DNA technology, scientists aim to engineer organisms with entirely new functions, such as bacteria that can produce biofuels or plants that can detect environmental pollutants. In medicine, personalized therapies are becoming possible, where recombinant DNA technology can be used to create tailored treatments based on an individual's genetic makeup. For instance, CAR-T cell therapy, a promising treatment for certain cancers, uses a patient's own cells, modified to better recognize and attack cancer cells. These advancements highlight the potential of recombinant DNA technology to enable more precise and effective treatments.

24.5 CONCLUSION

Recombinant DNA technology has had an undeniable impact on various fields, from medicine to agriculture and environmental science. Its ability to modify genetic material with precision offers solutions to complex challenges, whether through producing life-saving medicines, enhancing food security, or developing eco-friendly biotechnologies. However, as the field progresses, it is crucial to address ethical and safety concerns, ensuring responsible and beneficial applications. As technology advances, recombinant DNA will continue to shape our approach to health, the environment, and scientific research, driving innovation and opening new avenues for addressing pressing global issues.

PART 7:

CURRENT TRENDS AND NEW DIRECTIONS

CHAPTER 25

THE WIDER CONTRIBUTION OF MICROBIOLOGY TO THE PHARMACEUTICAL SCIENCES

Microbiology plays a foundational role in the pharmaceutical sciences, providing insights and techniques critical to the development, production, and quality assurance of medications. From understanding disease mechanisms to designing drugs, vaccines, and antimicrobial agents, microbiology has shaped pharmaceutical advances significantly. This scientific discipline focuses on the study of microorganisms, including bacteria, fungi, viruses, and parasites, which are key players in human health and disease. The application of microbiology in pharmaceuticals extends from drug discovery to the manufacture of sterile products, ensuring patient safety, therapeutic efficacy, and compliance with regulatory standards.

25.1 DRUG DISCOVERY AND DEVELOPMENT

Microbiology contributes extensively to drug discovery and development, particularly in the search for new antibiotics, antiviral agents, and vaccines. In the face of increasing drug-resistant infections, microbiologists are essential in identifying novel compounds with antimicrobial properties. Microbial organisms, like bacteria and fungi, produce bioactive molecules that can be used to treat various infections. Penicillin, the first widely used antibiotic, was discovered through the study of the *Penicillium* mold, a breakthrough that underscored the potential of microorganisms in therapeutic discovery. Today, the search for new antimicrobials often involves exploring less commonly studied environments, such as soil and marine ecosystems, to identify unique organisms and compounds that may lead to effective treatments.

In addition to antimicrobials, microbiology has a vital role in antiviral drug discovery. Researchers use viral cultures to screen potential antiviral agents, studying how drugs interact with viral replication processes. This approach led to the development of drugs for infections like HIV, influenza, and hepatitis. Vaccines, another major pharmaceutical achievement, rely on microbiological research to develop immunization strategies, leveraging attenuated or inactivated pathogens to create immunity against diseases.

25.2 QUALITY CONTROL AND MICROBIAL TESTING

Quality control is essential in pharmaceutical manufacturing to ensure products are safe, effective, and free from contamination. Microbiology provides various methods and tools for microbial testing, helping identify and control contamination throughout the production process. One of the primary tests used in quality control is the sterility test, which ensures that sterile products like injectable drugs, eye drops, and surgical materials are free from microorganisms. This test typically involves culturing samples under controlled conditions and observing them for any microbial growth.

Another important aspect of microbial testing is endotoxin testing, which detects bacterial endotoxins, substances that can cause severe immune reactions if introduced into the bloodstream. The Limulus Amebocyte Lysate (LAL) test, derived from horseshoe crab blood, is widely used to identify endotoxins in products like injectable drugs and intravenous fluids. Microbiologists also conduct bioburden testing to determine the number of viable microorganisms on or in a product. This testing helps monitor and control microbial presence, minimizing the risk of contamination and ensuring the safety of pharmaceutical products.

25.3 STERILIZATION TECHNIQUES AND ASEPTIC PROCESSING

Sterilization is a cornerstone of microbiology within pharmaceutical sciences, especially for products that must be free of any viable microorganisms. Various sterilization methods, including heat, filtration, radiation, and chemical agents, are employed based on the product's properties and microbial load. Autoclaving, a form of moist heat sterilization, is one of the most widely used methods, effectively killing bacteria, fungi, and viruses at high temperatures and pressure. For heat-sensitive materials, sterilization by filtration is used, where microorganisms are removed by passing the product through a fine filter.

Aseptic processing is another key aspect where microbiology contributes to pharmaceuticals, especially in the production of sterile products. Aseptic techniques prevent contamination during the manufacturing process, crucial for injectable drugs, biologics, and ophthalmic products. Microbiologists oversee aseptic procedures, ensuring that equipment, materials, and environments maintain the highest sterility standards. Cleanrooms and laminar flow hoods, essential for aseptic processing, are regularly monitored for microbial contamination, with protocols established for gowning, cleaning, and air filtration to maintain sterility.

25.4 DEVELOPMENT OF ANTIBIOTICS AND ANTIMICROBIAL RESISTANCE

The development of antibiotics is a field rooted deeply in microbiology. By studying microbial interactions, scientists have developed a range of antibiotics to combat bacterial infections. However, the overuse and misuse of antibiotics have led to antimicrobial resistance (AMR), a global health crisis. Microbiology's contribution to understanding AMR is invaluable in developing strategies to combat resistant infections. Through microbiological research, scientists investigate resistance mechanisms, such as gene mutations and efflux pumps, which bacteria use to evade antibiotic effects.

Microbiologists are also involved in surveillance programs that track the spread of resistance, identifying patterns and informing guidelines on antibiotic usage. In pharmaceutical development, microbiology helps identify new antibiotic targets and design molecules that can bypass existing resistance mechanisms. Furthermore, microbiology provides insights into alternative approaches, like bacteriophage therapy, which uses viruses to target specific bacteria, potentially offering a solution to drug-resistant infections.

25.5 BIOPHARMACEUTICALS AND BIOTECHNOLOGY

Biopharmaceuticals, or biologics, are a category of drugs derived from biological sources, including microorganisms. Microbiology plays a critical role in biopharmaceutical production, particularly in the use of genetically engineered bacteria, yeasts, and cell lines to produce therapeutic proteins, enzymes, and monoclonal antibodies. For example, insulin, once extracted from animal sources, is now produced by genetically modified *Escherichia coli* or yeast, ensuring a more consistent and safe product for diabetic patients.

Recombinant DNA technology, a microbiological advancement, enables scientists to insert human genes into microorganisms, which then produce human proteins that can be used in therapies for diseases like cancer, rheumatoid arthritis, and multiple sclerosis. Microbiologists also ensure the quality of biopharmaceuticals by monitoring microbial contamination, evaluating the purity of products, and testing for potential immunogenic reactions.

25.6 VACCINE DEVELOPMENT AND IMMUNIZATION

Vaccines are one of the most impactful contributions of microbiology to human health, providing immunity against numerous infectious diseases. Microbiology is at the heart of vaccine development, from studying pathogens and their life cycles to designing and testing vaccines that provide immunity without causing disease. Traditional vaccines use live attenuated or inactivated pathogens, while modern vaccines, such as mRNA vaccines, rely on genetic material from pathogens to stimulate an immune response. Microbiologists also contribute to the testing and quality control of vaccines, ensuring they are safe, effective, and meet regulatory standards. Vaccine development involves rigorous preclinical and clinical testing phases, where microbiologists assess the immune response, safety, and efficacy of vaccine candidates. Their work ensures that vaccines protect against diseases while minimizing side effects and potential risks.

25.7 ENVIRONMENTAL MICROBIOLOGY IN PHARMACEUTICAL PRODUCTION

Environmental microbiology addresses the impact of environmental microorganisms on pharmaceutical production. Microbial contamination can originate from various sources, including air, water, and raw materials used in production. Environmental monitoring is therefore critical to maintaining a contamination-free manufacturing environment. Microbiologists are responsible for testing air and surfaces in production areas, especially in aseptic facilities, to detect any contamination. Water quality is also essential in pharmaceutical production, as water is used in numerous processes and formulations. Microbiologists ensure that water meets microbiological standards by testing for bacteria, endotoxins, and other contaminants. By understanding and controlling environmental microorganisms, microbiology helps safeguard the quality of pharmaceutical products.

25.8 FUTURE DIRECTIONS AND INNOVATIONS IN MICROBIOLOGY AND PHARMACEUTICALS

As new challenges emerge in healthcare, the role of microbiology in the pharmaceutical sciences continues to grow. Advances in genomics and bioinformatics allow microbiologists to study pathogens at the genetic level, offering new insights into disease mechanisms and potential therapeutic targets. Rapid diagnostic tools, like polymerase chain reaction (PCR) and next-generation sequencing, enable faster identification of pathogens and resistance genes, crucial for timely and effective treatment decisions. Emerging fields like synthetic biology are also transforming microbiology in pharmaceuticals, enabling the design of custom microorganisms to produce complex drugs or break down harmful chemicals in the environment. Additionally, the development of probiotics and microbiome-based therapies shows promise in treating diseases linked to gut health and immunity.

25.9 CONCLUSION

Microbiology has a profound and wide-reaching impact on the pharmaceutical sciences, driving innovations from drug discovery to quality control and vaccine development. Through an understanding of microbial life, the pharmaceutical industry continues to evolve, addressing critical health challenges and improving the safety, efficacy, and accessibility of treatments. As microbiology advances, so too does its contribution to pharmaceutical sciences, offering solutions to new and existing medical challenges and ensuring that therapeutic products are produced to the highest standards of safety and quality.

CHAPTER 26

ALTERNATIVE STRATEGIES OF ANTIBIOTICS: PRIORITIES FOR DEVELOPMENT

Antibiotic resistance poses a significant challenge to global health, as more pathogens develop immunity to traditional treatments. The emergence of multidrug-resistant (MDR) bacteria has spurred a global need to find alternative strategies to conventional antibiotics. Such alternatives aim to prevent, manage, or treat infections in ways that do not promote resistance or rely solely on traditional antibiotics.

Research into these alternatives has led to promising new methods, including bacteriophage therapy, antimicrobial peptides, probiotics, immune-modulation therapies, and anti-virulence strategies. Each of these methods targets microbial activity or infection processes in novel ways, offering avenues for effective treatments and sustainable control of resistant bacteria.

26.1 BACTERIOPHAGE THERAPY

One of the most promising alternative strategies to traditional antibiotics is bacteriophage therapy, which utilizes viruses known as bacteriophages to target and destroy specific bacterial pathogens. Bacteriophages, or phages, are naturally occurring viruses that infect bacteria. When they invade bacterial cells, they replicate and ultimately cause the bacterial cell to lyse, releasing new phages that continue to attack additional bacteria.

Unlike broad-spectrum antibiotics, bacteriophages are highly specific, often targeting only a single species or strain of bacteria. This specificity minimizes damage to beneficial microbiota and reduces the likelihood of side effects. Additionally, bacteriophages can evolve alongside bacteria, potentially allowing them to keep pace with bacterial resistance. Despite its promise, bacteriophage therapy faces challenges, including regulatory hurdles, variability in phage effectiveness, and issues related to the development of resistance against phages themselves. However, ongoing research aims to optimize bacteriophage therapies and integrate them into clinical settings as a complementary or alternative treatment to antibiotics.

26.2 ANTIMICROBIAL PEPTIDES (AMPS)

Antimicrobial peptides (AMPs) are another promising alternative to conventional antibiotics. AMPs are short chains of amino acids that exhibit potent antimicrobial properties, targeting a wide range of pathogens, including bacteria, fungi, and viruses. Many organisms, from humans to insects, produce these peptides as a part of their innate immune defense. AMPs operate by disrupting microbial cell membranes or by interfering with essential cellular processes within pathogens.

One of the critical advantages of AMPs is their ability to act quickly, often killing bacteria within minutes. Furthermore, because they target the bacterial cell membrane rather than specific internal pathways, bacteria are less likely to develop resistance to AMPs. However, the therapeutic application of AMPs is still under investigation, as some peptides can be toxic to human cells or degrade rapidly in the body. Researchers are working to modify and stabilize AMPs, ensuring that they are safe, effective, and long-lasting in clinical applications.

26.3 PROBIOTICS AND MICROBIOME MODULATION

Probiotics and microbiome modulation represent a natural approach to combat pathogenic bacteria by promoting beneficial bacteria. Probiotics are live microorganisms that, when administered in adequate amounts, provide health benefits, particularly by enhancing the immune system and competing with harmful bacteria for resources and space. Certain probiotic strains can inhibit pathogenic bacteria through the production of substances like bacteriocins and organic acids, which inhibit the growth of harmful microbes.

Modulating the microbiome by promoting beneficial bacteria helps maintain a balanced microbial environment, preventing infections and reducing the need for antibiotics. This approach is particularly relevant for infections related to disruptions in the natural microbiome, such as gastrointestinal infections. While promising, probiotics and microbiome modulation therapies require more research to understand their full capabilities, mechanisms of action, and optimal delivery methods in treating specific infections.

26.4 IMMUNE-MODULATION THERAPIES

Enhancing the body's immune response to combat infections offers an alternative strategy that bypasses the need for antibiotics. Immune-modulation therapies involve the use of agents that stimulate or modulate the immune system to fight infections more effectively. By strengthening the immune response, the body can better prevent or eliminate bacterial infections without relying on antibiotics.

One example of immune-modulation therapy is the use of cytokines, which are signaling proteins that regulate immune responses. Administering cytokines such as interleukins can boost the immune system's ability to recognize and destroy pathogens. Additionally, treatments like vaccines prime the immune system against specific pathogens, reducing the need for antibiotics. Immunotherapies are promising but must be carefully regulated to avoid over-stimulating the immune response, which can lead to harmful side effects. Nevertheless, as research advances, immune-modulation therapies could become essential tools in the fight against antibiotic-resistant infections.

26.5 ANTI-VIRULENCE STRATEGIES

Anti-virulence strategies aim to neutralize the pathogenicity of bacteria without killing them outright. Unlike traditional antibiotics that eliminate bacteria, anti-virulence agents target specific mechanisms that pathogens use to cause disease, such as toxins, adhesion factors, or communication systems known as quorum sensing. By disabling these virulence factors, bacteria lose their ability to cause infection, allowing the body's immune system to clear the infection naturally.

The advantage of anti-virulence therapies is that they exert less selective pressure on bacteria, which reduces the likelihood of resistance development. This approach also allows beneficial bacteria to remain unaffected, maintaining a balanced microbiome. However, developing anti-virulence agents is complex, as it requires detailed knowledge of bacterial virulence mechanisms and the ability to target them specifically. Currently, anti-virulence therapies are still under investigation, with some promising compounds showing efficacy in preclinical studies.

26.6 CRISPR-CAS TECHNOLOGY

CRISPR-Cas technology, widely recognized for its potential in gene editing, also offers exciting possibilities as an alternative to antibiotics. CRISPR systems can be programmed to target and cut specific DNA sequences within bacterial genomes, effectively eliminating harmful bacteria. This precision allows scientists to selectively target pathogenic bacteria without affecting beneficial ones, which is a significant advantage over broad-spectrum antibiotics.

CRISPR technology has the potential to eradicate bacteria with specific resistance genes, effectively "editing out" resistance traits and preventing the spread of resistance within microbial communities. Although CRISPR as an antimicrobial tool is still in its early stages, its specificity and adaptability make it a promising candidate for future therapies. Challenges remain in delivery mechanisms and potential immune responses to CRISPR-based therapies, but the technology holds substantial potential in combating drug-resistant infections.

26.7 NATURAL COMPOUNDS AND PLANT EXTRACTS

The use of natural compounds and plant extracts as antimicrobial agents is a traditional approach that has gained renewed interest as an alternative to antibiotics. Many plants produce secondary metabolites with antimicrobial properties, such as alkaloids, flavonoids, and essential oils. These compounds can inhibit microbial growth, disrupt cell membranes, or interfere with microbial enzymes.

For example, compounds from garlic, turmeric, and tea tree oil have demonstrated significant antimicrobial effects against various pathogens. The advantage of natural compounds is that they often have multiple modes of action, making it more difficult for bacteria to develop resistance. However, more research is needed to identify the most effective compounds, determine optimal dosages, and develop standardized formulations for clinical use. Nonetheless, natural compounds provide a rich source of potential alternatives to antibiotics, particularly for topical applications and supportive therapies.

26.8 NANOTECHNOLOGY-BASED ANTIMICROBIALS

Nanotechnology offers innovative approaches to developing alternative antimicrobial agents. Nanoparticles, such as silver, gold, and zinc oxide, have shown promising antimicrobial properties. These nanoparticles can penetrate bacterial cells, disrupt cellular processes, and generate reactive oxygen species that lead to cell death. Their small size allows them to interact with bacterial cells in unique ways, potentially overcoming resistance mechanisms. Moreover, nanotechnology enables targeted drug delivery, allowing antimicrobials to reach infection sites with higher precision and reduced side effects. While the clinical use of nanoparticles in antimicrobials is still emerging, research indicates that nanotechnology has the potential to revolutionize the field. Safety concerns regarding toxicity and the environmental impact of nanoparticles remain, but nanotechnology-based antimicrobials may become valuable alternatives in the future.

26.9 CONCLUSION

The development of alternative strategies to antibiotics is critical to addressing the growing crisis of antimicrobial resistance. Approaches such as bacteriophage therapy, antimicrobial peptides, probiotics, immune-modulation therapies, and anti-virulence strategies offer diverse mechanisms to combat infections without relying solely on traditional antibiotics. Each strategy has unique advantages, from targeted action to minimized resistance development. However, significant research, regulatory approval, and technological advances are required to bring these alternatives into widespread clinical use. By investing in these promising alternatives, the pharmaceutical and healthcare industries can better manage infectious diseases and sustain effective infection control in the face of evolving resistance.

BIBLIOGRAPHY

Biopharmaceuticals: Biochemistry and Biotechnology by J. R. G. O'Connor Publication Year: 2010

Biotechnology and Pharmaceutical Microbiology by D. P. A. Kumar Publication Year: 2005

Introduction to Pharmaceutical Microbiology by R. J. Tompkins Publication Year: 2011

Microbial Contamination Control in Parenteral Manufacturing by R. F. Shaffer Publication Year: 2011

Microbial Control in Pharmaceutical Manufacturing by G. J. Brown Publication Year: 2017

Microbial Infections in Pharmaceutical Industries by M. G. Smiley Publication Year: 2000

Microbial Infections in Pharmaceutical Products and Drugs by T. M. Weathers Publication Year: 2009

Microbial Techniques in Pharmaceutical Manufacturing by S. A. Malik Publication Year: 2014

Microbial Testing in the Pharmaceutical Industry by R. P. Singh Publication Year: 2017

Microbiological Applications in Pharmaceutical Manufacturing by K. K. Gupta Publication Year: 2010

Microbiology and Pharmacology: An Integrated Approach by S. A. Johnson Publication Year: 2012

Microbiology for Pharmaceutical Sciences: Fundamentals and Applications by G. V. Mohanty Publication Year: 2008

Microbiology for Pharmacy Students by D. H. D. P. Tewari Publication Year: 2015

Microbiology for the Pharmaceutical Sciences by W. L. Finkelstein Publication Year: 2000

Microbiology in Pharmaceutical Development by G. L. Silva Publication Year: 2003

Microbiology of Pharmaceutical Products by B. K. Mishra Publication Year: 2014

Modern Pharmaceutical Microbiology by P. S. Murthy Publication Year: 2009

Pharmaceutical Biotechnology and Microbiology by G. H. Kohli Publication Year: 2016

Pharmaceutical Microbiology and Biotechnology by T. R. Roy Publication Year: 2013

Pharmaceutical Microbiology and Biotechnology by W. G. V. P. Mathews Publication Year: 2018

Pharmaceutical Microbiology and Microbial Techniques by P. A. Dewitt Publication Year: 2015

Pharmaceutical Microbiology by L. S. M. Sabouraud Publication Year: 1964

Pharmaceutical Microbiology Handbook by J. A. Palmer Publication Year: 2008

Pharmaceutical Microbiology in Industry by P. J. Houghton Publication Year: 2015

Pharmaceutical Microbiology in Practice by M. H. Krishnan Publication Year: 2014

Pharmaceutical Microbiology: A Guide for Students and Practitioners by R. G. G. Russell Publication Year: 2002

Pharmaceutical Microbiology: A Practical Approach by A. L. P. Brown Publication Year: 2009

Pharmaceutical Microbiology: Antimicrobial Agents and Antimicrobial Resistance by R. S. R. Smith Publication Year: 2011

Pharmaceutical Microbiology: Applications in Quality Control by N. P. Jadhav Publication Year: 2017

Pharmaceutical Microbiology: Concepts and Practices in Manufacturing by R. P. Thakur Publication Year: 2017

Pharmaceutical Microbiology: Concepts, Issues, and Applications by M. G. Thurston Publication Year: 2011

Pharmaceutical Microbiology: Current Research by N. R. Cheng Publication Year: 2020

Pharmaceutical Microbiology: Fundamentals and Clinical Applications by D. K. Lee Publication Year: 2018

Pharmaceutical Microbiology: Fundamentals by P. A. White Publication Year: 2012

Pharmaceutical Microbiology: Molecular Approaches and Practices by D. H. James Publication Year: 2016

Pharmaceutical Microbiology: Practices and Procedures by S. P. Tyagi Publication Year: 2016

Pharmaceutical Microbiology: Principles and Applications by K. M. Patel Publication Year: 2013

Pharmaceutical Microbiology: Principles and Practice in Manufacturing by C. D. E. Whelan Publication Year: 2011

Pharmaceutical Microbiology: Quality Assurance and Control by N. C. Rao Publication Year: 2013

Pharmaceutical Microbiology: Recent Trends and Innovations by M. G. Enomoto Publication Year: 2020

Pharmaceutical Microbiology: Regulation, Quality Assurance, and Validation by G. T. Bhat Publication Year: 2014

Pharmaceutical Microbiology: Research and Practice by R. E. Bates Publication Year: 2012

Pharmaceutical Microbiology: Techniques and Applications by S. R. Suryawanshi Publication Year: 2015

Pharmaceutical Microbiology: Theory and Practice by G. S. Bansal Publication Year: 2015

Pharmaceutical Microbiology: Understanding the Basics by S. K. Mohapatra Publication Year: 2012

Principles of Pharmaceutical Microbiology by A. S. K. Emsley Publication Year: 2016

Principles of Pharmaceutical Microbiology by J. A. McCutcheon Publication Year: 2014

Sterile Drug Products: Manufacturing, Quality Control, and Regulation by M. M. Reitz Publication Year: 2017

The Basics of Pharmaceutical Microbiology by K. S. S. P. Surya Publication Year: 2015

The Role of Microbiology in Pharmaceutical Development by J. A. Smith Publication Year: 2007

INDEX

H

I

J

K

L

M

N

O

P

Q

R

S

www.ingramcontent.com/pod-product-compliance
Lightning Source LLC
LaVergne TN
LVHW062219200926
848484LV00006B/1431

* 9 7 8 1 7 8 7 1 5 4 4 6 9 *